The ASQ Certified Six Sigma Black Belt Study Guide

The ASQ Certified Six Sigma Black Belt Study Guide

Second Edition

Mary McShane-Vaughn, Editor

ASQExcellence
Milwaukee, Wisconsin

The ASQ Certified Six Sigma Black Belt Study Guide, Second Edition
Published by ASQExcellence
Produced and distributed by Quality Press, ASQ, Milwaukee, WI

Publisher's Cataloging-in-Publication data

Names: McShane-Vaughn, Mary, editor.
Title: The ASQ certified Six Sigma black belt study guide , second edition / Mary McShane-Vaughn, Editor.
Description: Includes bibliographical references. | Milwaukee, WI: ASQExcellence, 2023.
Identifiers: LCCN: 2023944352 | ISBN: 978-1-63694-116-5 (paperback) | 978-1-63694-117-2 (PDF) | 978-1-63694-118-9 (epub)
Subjects: LCSH Six sigma (Quality control standard)—Study guides. | Production management—Study guides. | Quality control—Statistical methods—Study guides. | BISAC BUSINESS & ECONOMICS / Quality Control | STUDY AIDS / Professional
Classification: LCC TS155 .A77 2023 | DDC 658.5/62—dc23

ASQ and ASQExcellence advance individual, organizational, and community excellence worldwide through learning, quality improvement, and knowledge exchange.

Bookstores, wholesalers, schools, libraries, businesses, and organizations: Quality Press and ASQExcellence books are available at quantity discounts for bulk purchases for business, trade, or educational uses. For more information, please contact Quality Press at 800-248-1946 or books@asq.org.

To place orders or browse the selection of all Quality Press and ASQExcellence titles, visit our website at: http://www.asq.org/quality-press.

27 26 25 24 23 LS 6 5 4 3 2 1

ASQ
Quality Press
600 N. Plankinton Ave.
Milwaukee, WI 53203-2914
Email: books@asq.org
Excellence Through Quality™

ASQE™
Inspire Excellence

For Jim, my favorite study partner.

Table of Contents

Preface

This study guide is a valuable resource for candidates preparing for the ASQ Certified Six Sigma Black Belt (CSSBB) exam and is best used in combination with *The ASQ Certified Six Sigma Black Belt Handbook, Fourth Edition* (ASQExcellence).

The book is divided into two parts that provide a thorough review of the 2022 Six Sigma Black Belt Body of Knowledge (BoK). Part A contains 550 multiple-choice practice questions and answers organized by BoK section. Part B presents 150 additional questions and answers spanning the entire BoK in random order to simulate an actual certification exam. For each solution, the specific area of the BoK being tested is listed so that candidates can focus their study on sections that they find difficult.

Unlike other resources on the market, all the questions and solutions were developed specifically to address the 2022 CSSBB BoK, including considering the proper depth of knowledge and required levels of cognition.

To get the most from this study guide, solutions for the mathematical questions should be worked out using a calculator from the ASQ-approved list. In addition, *The ASQ Certified Six Sigma Black Belt Handbook* from Quality Press is the best overall reference for the latest BoK. Using the handbook to answer questions will help candidates become familiar with the needed equations and statistical tables before they sit for the exam.

The Black Belt certification exam is open-book. It is strongly recommended that you take *The ASQ Certified Six Sigma Black Belt Handbook* into the exam to look up or verify answers. As a reminder, any sample or practice questions such as those in this study guide cannot be taken into ASQ certification exam rooms.

I'd like to thank Michael Byrnes, Executive Director of Certification Operations at ASQExcellence, for giving me the opportunity to work with Lillian McAnally and Erica Barse of Quality Press once again. In addition, thanks to Dr. Stephen P. Bailey, Dr. Courtney Chronley, Dr. Sandra Furterer, Mary Glas CSSBB, Dr. Karen Hulting, and Mary Beth Soloy CMBB for reviewing the questions for clarity and the solutions for accuracy. It was a daunting task!

I welcome your feedback and suggestions for improvement. Please contact me at books@asq.org, and I will do my best to clarify questions you may have and incorporate any suggestions for improvement into future printings or editions of this study guide.

About the Author

Dr. Mary McShane-Vaughn is the founder of University Training Partners, a company that develops and delivers online Lean Six Sigma and statistics training. Over the past 13 years, she has enthusiastically trained over 1400 Six Sigma Yellow, Green, and Black Belt students.

Mary is the editor of the latest edition of *The ASQ Certified Six Sigma Black Belt Handbook, Fourth Edition* which reflects the updated 2022 Body of Knowledge. She is also the author of three textbooks: *The Probability Handbook, The Probability Workbook,* and *Lean Six Sigma Leadership Tools for Black Belts,* and a coauthor of *The Handbook of Quality Equations* and *The Certified Quality Inspector Handbook.*

Dr. McShane-Vaughn holds ASQ certifications as a Six Sigma Black Belt, Quality Engineer, and Reliability Engineer. She is a past ASQ Black Belt exam chair and the current quality engineer exam chair. She also serves on the ASQ Technical Program committee.

Mary earned her PhD in industrial engineering and an MS in statistics from the Georgia Institute of Technology, and a BS in industrial engineering from Kettering University in Michigan.

Limit of Liability/Disclaimer of Warranty

The author has put forth their best effort in compiling the content of this book; however, no warranty with respect to the material's accuracy or completeness is made. Additionally, no warranty is made in regard to applying the recommendations made in this book to any business structure or environment. Businesses should consult regulatory, quality, and/or legal professionals prior to deciding on the appropriateness of advice and recommendations made within this book. The author shall not be held liable for loss of profit or other commercial damages resulting from the employment of recommendations made within this book, including special, incidental, consequential, or other damages.

Part A
Practice Questions by 2022 BoK Section

In this first part of the study guide, questions and answers are presented in order of the nine sections of the 2022 Body of Knowledge. The number of practice questions included in each BoK section is shown in the table below.

BoK Section	Topic	Number of Practice Questions
Section I	Organization-Wide Planning and Deployment	37
Section II	Organizational Process Management and Measures	38
Section III	Team Management	52
Section IV	Define	56
Section V	Measure	151
Section VI	Analyze	68
Section VII	Improve	63
Section VIII	Control	65
Section IX	Design for Six Sigma (DFSS) Framework and Methodologies	20

Section I

Organization-Wide Planning and Deployment

I. Organization-Wide Planning and Deployment (12 Questions)

A. *Organization-wide considerations*

1. *Fundamentals of six sigma and lean methodologies.* Define and describe the value and goals of these approaches and describe the integration and complementary relationship between them. Identify and understand an organization's lean six sigma maturity model. (Understand)

2. *Six sigma, lean, and continuous improvement methodologies.* Describe when to use six sigma instead of other problem-solving approaches and describe the importance of aligning six sigma objectives with organizational goals. Describe screening criteria and how such criteria can be used for the selection of six sigma projects, lean initiatives, and other continuous improvement methods. (Apply)

3. *Relationships among business systems and processes.* Describe the interactive relationships among business systems, processes, and internal and external stakeholders, and the impact those relationships have on business systems. (Understand)

4. *Strategic planning and deployment of initiatives.* Define the importance of identification and strategic planning of six sigma projects and lean initiatives as part of the business planning process taking into consideration outcomes such as return on investment and measured intangibles. Use strengths, weaknesses, opportunities, and threats analysis (SWOT), contingency planning, and business continuity planning to enhance strategic planning and deployment. (Apply)

B. *Leadership*

1. *Roles and responsibilities.* Describe the roles and responsibilities of executive leadership, champions, sponsors, process owners, master black belts, black belts, and green belts in driving six sigma and lean initiatives. Understand the importance of coaching multiple levels of leadership and belts on appropriate six sigma tools and techniques. Understand the importance of finance and its role supporting a project and confirming its outcome.

Describe how each group influences project deployment in terms of providing or managing resources, enabling changes in organizational structure, and supporting communications about the purpose and deployment of the initiatives. (Understand)

2. *Organizational barriers*. Describe how an organization's structure and culture can impact six sigma projects. Identify common causes of six sigma failures, including lack of management support and lack of resources. (Apply)

3. *Change management*. Apply and facilitate change management techniques, including stakeholder analysis, readiness assessments, proactive change management, and communication plans to overcome barriers and drive organization-wide change. (Apply)

QUESTIONS

1. A common thread in the use of the Six Sigma approach across organizations is the:
   a. separation of Six Sigma and business strategy.
   b. minimal use of statistical techniques.
   c. emphasis on quick-turnaround projects that have the highest payoff.
   d. use of teams assigned to well-defined projects that impact the bottom line.
2. In Six Sigma, the concept of $Y = f(X)$ refers to:
   a. the relationship between outputs and customer satisfaction.
   b. how inputs and the process affect outputs.
   c. the connection between statistics and process improvement.
   d. how quality transforms organizations.
3. Categories that are typically evaluated in a Six Sigma maturity model include:
   a. number of employees trained and number of concurrent projects.
   b. leadership, customer focus, and business results.
   c. number of statistical software licenses purchased and black belts trained.
   d. stock price and reductions in staff.
4. Six Sigma can be thought of as:
   a. a philosophy, a methodology, and a statistical standard of quality.
   b. a method that uses new tools to solve process problems.
   c. a project approach that uses six steps to reduce defects.
   d. a quality method that emphasizes reducing waste.
5. One of the five pillars, or principles, of lean is:
   a. reducing the risk of external audit findings.
   b. aiming for 3.4 defects per million opportunities.
   c. identifying the value from the customers' point of view.
   d. creating organized warehouses to store excess inventory.

6. A lean organization is characterized by employees who:
   a. are well-informed and cross-trained.
   b. are statistical experts.
   c. are fearful of change.
   d. use a five-step improvement process to solve problems.
7. Which of the following terms is associated with the third stage in a Six Sigma maturity model?
   a. Launch
   b. Culture transformation
   c. Scale and replication
   d. Institutionalization
8. During the early-success phase of a Six Sigma implementation:
   a. all belts have undergone training.
   b. projects are geared toward low-hanging fruit.
   c. DMAIC is consistently used to solve problems.
   d. Six Sigma is ingrained into the organization's culture.
9. Six Sigma projects should:
   a. use all seven of the basic quality tools.
   b. involve a designed experiment.
   c. be completed within two months.
   d. involve frontline workers affected by the project.
10. How should Six Sigma projects be selected?
    a. Based on criteria such as customer impact and availability of resources
    b. According to the opinions of top management
    c. Only if there is a known solution
    d. Based on public relations feedback
11. Using lean and Six Sigma together:
    a. reduces variation and waste.
    b. only succeeds in manufacturing environments.

c. creates an excessive training burden.

d. fosters an "us" versus "them" working environment.

12. What is the relationship between a process and a system?

a. Processes are system outputs.

b. Systems are a series of interrelated processes.

c. Systems are process inputs.

d. Processes are a series of interrelated systems.

13. Which of the following groups is an example of an internal stakeholder?

a. Regulatory agencies

b. R&D department

c. Shareholders

d. Suppliers

14. Which of the following quality tools is used to diagram a process?

a. Control chart

b. Measles chart

c. Flowchart

d. House of quality

15. Which of the following tools would be the most appropriate for crisis management?

a. SWOT analysis

b. SMART goals

c. Contingency planning

d. Risk analysis

16. Which of the following categories in a SWOT analysis are considered external?

a. Opportunities and strengths

b. Weaknesses and threats

c. Opportunities and threats

d. Strengths and weaknesses

17. Issues that a business continuity plan may address include:

    a. product recalls.

    b. loss of key personnel.

    c. pandemics.

    d. loss of data.

18. Six Sigma projects can deliver nonfinancial benefits, such as increased customer satisfaction. These benefits are referred to as:

    a. measured intangibles.

    b. false savings.

    c. imaginary outcomes.

    d. unmeasurable benefits.

19. Which of the following conditions might be addressed in a contingency plan for asset security?

    a. Loss of key personnel

    b. Pandemics

    c. Product recalls

    d. On-the-job accidents

20. Six Sigma projects may deliver soft savings such as:

    a. increased customer satisfaction.

    b. reduced variation.

    c. increased profits.

    d. decreased rework expenses.

21. The term *champion* is usually associated with an individual who:

    a. ensures that their projects are aligned with the organization's strategic goals and priorities, and removes organizational barriers.

    b. retains their regular position but is trained in the tools, methods, and skills necessary to conduct Six Sigma improvement projects.

    c. is typically assigned full-time to train, mentor, and lead improvement projects.

    d. is typically assigned full-time to train, mentor, and lead the strategy for chartering organizations' strategic projects.

22. A process owner is an individual who:

    a. retains their regular position, but is trained in the tools, methods, and skills necessary to conduct Six Sigma improvement projects.

    b. is typically assigned full-time to train, mentor, and lead improvement projects.

    c. is typically assigned full-time to train, mentor, and lead the strategy for chartering organizations' strategic projects.

    d. has the authority or ability to make changes in the process as required.

23. For a Six Sigma project, which action might the finance group perform?

    a. Selecting projects based on return on investment

    b. Vetting the estimated savings for a project

    c. Denying capital expenditures

    d. Developing a workforce reduction plan

24. The term *Green Belt* is usually associated with an individual who:

    a. retains their regular position but is trained in the tools, methods, and skills necessary to conduct Six Sigma improvement projects.

    b. is typically assigned full-time to train, mentor, and lead improvement projects.

    c. is typically assigned full-time to train, mentor, and lead the strategy for chartering organizations' strategic projects.

    d. has the authority or ability to make changes in the process as required.

25. Which Six Sigma role is called upon to remove roadblocks when implementing change?

    a. Black Belt

    b. Master Black Belt

    c. Manager

    d. Champion

26. The term *Black Belt* is usually associated with an individual who:

    a. is typically assigned full-time to train, mentor, and lead the strategy for chartering organizations' strategic projects.

    b. retains their regular position but is trained in the tools, methods, and skills necessary to conduct Six Sigma improvement projects.

c. is typically assigned full-time to train, mentor, and lead improvement projects and has the authority or ability to make changes in the process as required.

d. is typically assigned full-time to train, mentor, and lead the strategy for chartering the organizations' strategic projects.

27. Which Six Sigma role is involved in giving on-the-spot training, advising team leaders, and helping a team prepare for tollgate meetings?

   a. Executive

   b. Yellow belt

   c. Finance

   d. Coach

28. The term *Master Black Belt* is usually associated with an individual who:

   a. retains their regular position but is trained in the tools, methods, and skills necessary to conduct Six Sigma improvement projects.

   b. is typically assigned full-time to train, mentor, and lead improvement projects.

   c. is typically assigned full-time to train, mentor, and lead the strategy for chartering organizations' strategic projects.

   d. has the authority or ability to make changes in the process as required.

29. Kurt Lewin proposed a change management theory that includes which three stages?

   a. Define, manage, and communicate

   b. Unfreeze, change, and freeze

   c. Lock, communicate, and implement

   d. Plan, change, and act

30. Successful Six Sigma implementations:

   a. are driven by the frontline workers.

   b. are driven from the top-down, with executives serving as Master Black Belts.

   c. involve a steering committee with top management representation and support.

   d. are rarely successful in their first year.

31. A critical success factor for a Six Sigma implementation is:
    a. communication.
    b. advanced statistical expertise.
    c. requiring Black Belts to complete four projects per year.
    d. financial rewards for participants.

32. Which tool would be the best to use to assess an organization's willingness to change its culture?
    a. SWOT analysis
    b. Readiness assessment
    c. Interrelationship digraph
    d. Stakeholder analysis

33. When performing a force-field analysis, the team lists a desired state and then brainstorms:
    a. causes and effects.
    b. strengths and weaknesses.
    c. driving and restraining forces.
    d. selection criteria.

34. Effective communication is essential to successfully implementing change. Which of the following items is typically communicated first?
    a. A view of the future state
    b. Goals and metrics
    c. New policies
    d. The need for change

35. Which of the following types of organizational structure is most resistant to change?
    a. Matrix
    b. Cross-functional
    c. Centralized
    d. Decentralized

36. Which of the following models is the best for identifying stakeholders of a proposed change in a manufacturing process?

    a. SIPOC

    b. PDCA

    c. DMAIC

    d. DFSS

37. To maintain support throughout the change life cycle, the organization must:

    a. provide frequent communications.

    b. appoint a management representative.

    c. delegate the responsibilities to a qualified consultant.

    d. divide the responsibility and authority equally between management and the workers.

SOLUTIONS

1. d; The Six Sigma approach emphasizes assigning teams to tackle well-defined projects that bring financial reward to the organization. The projects should mesh with the organization's business strategy and can run anywhere from two to nine months. Decisions are made based on data, and statistical techniques are often used for analysis. [I.A.1]

2. b; In Six Sigma, the concept of $Y = f(X)$ refers to a relationship in which the Y variable represents outputs and X represents the inputs. The function f represents the process that transforms the inputs into the outputs. By understanding this relationship, an organization can predict and control process outputs based on its inputs. [I.A.1]

3. b; Typical categories addressed in a Six Sigma maturity model include Six Sigma leadership, customer focus, Six Sigma strategy, Six Sigma project management, evaluation and motivation, Six Sigma infrastructure, and business results. [I.A.1]

4. a; Six Sigma is a philosophy, an improvement methodology that uses five project phases (DMAIC), and a statistical standard of quality (3.4 DPMO). Benefits of the Six Sigma methodology include reduced variation, decreased defect levels, and increased throughput. The lean approach, in contrast, emphasizes reducing waste. [I.A.1]

5. c; Identifying the value from the customers' point of view is a pillar of the lean approach. The five pillars are (1) specify value from the customers' point of view, (2) identify the value stream, (3) create flow, (4) pull from the customer, (5) seek perfection. [I.A.1]

6. a; A lean organization is characterized by employees who are well-informed and cross-trained and who participate directly in improving processes. [I.A.1]

7. c; Scale and replication. The five stages of Six Sigma maturity are (1) launch, (2) early success, (3) scale and replication, (4) institutionalization, (5) culture transformation. [I.A.1]

8. b; In the early success phase of Six Sigma implementation, projects are geared toward low-hanging fruit. These early successes demonstrate proof of concept and prepare the organization for the scale and replication phase. [I.A.1]

9. d; Six Sigma projects should involve the frontline workers of the process being improved. These workers are stakeholders in the project and have key insights into the process. Their support is needed for any improvements to stick. [I.A.2]

10. a; Six Sigma projects should be chosen based on an agreed-upon set of weighted criteria, such as customer impact and availability of resources, among others. [I.A.2]

11. a; Using lean and Six Sigma together reduces variation and waste. Six Sigma concentrates on reduced variation, while lean emphasizes elimination of waste. [I.A.2]

12. b; A system is a group of interdependent processes and people that together perform a common mission. [I.A.3]

13. b; The R&D department is internal to the company. All other entities listed are possible stakeholders but are external to the organization. [I.A.3]

14. c; Flowcharts are graphical representations of the steps in a process and are drawn to better understand processes. [I.A.3]

15. c; Contingency planning (also called a *plan B*) is used for crisis management, business continuity, and asset security. [I.A.4]

16. c; Opportunities and threats are considered external to the company's direct control. Strengths and weaknesses are identified with respect to the internal capabilities of an organization, while opportunities and threats look outside the organization to identify opportunities for the organization and threats to the organization. [I.A.4]

17. b; A business continuity plan may address issues such as loss of key personnel, a sudden change in the marketplace, or a strike by a bargaining unit. [I.A.4]

18. a; Nonfinancial benefits or soft savings such as increased customer satisfaction are referred to as measured intangibles. For example, customer satisfaction can be measured via net promoter scores, market share, or returns reports. [I.A.4]

19. c; All of the options may be addressed in contingency plans, but product recalls would be addressed in an asset security contingency plan. [I.A.4]

20. a; Increased customer satisfaction is considered soft savings. The other project outcomes are linked to hard dollar savings. [I.A.4]

21. a; Champion is the Six Sigma role associated with a senior manager who ensures that his or her projects are aligned with the organization's strategic goals and priorities, provides the Six Sigma team with resources, removes organizational barriers for the team, participates in project tollgate reviews, and essentially serves as the team's backer. A champion is also known as a sponsor. [I.B.1]

22. d; Process owner is the Six Sigma role associated with an individual who coordinates the various functions and work activities at all levels of a process, has the authority or ability to make changes in the process as required, and manages the entire process cycle so as to ensure performance effectiveness. [I.B.1]

23. b; The finance department contributes to the Six Sigma effort by vetting the estimated savings for a project. [I.B.1]

24. a; A Green Belt (GB) generally retains their regular position but is trained in the tools, methods, and skills necessary to conduct Six Sigma improvement projects either individually or as part of larger teams. [I.B.1]

25. d; A project champion is an upper-level manager who serves as a sounding board and can work to remove roadblocks that the improvement team encounters. [I.B.1]

26. c; A Black Belt (BB) is a Six Sigma role associated with an individual typically assigned full-time to train and mentor Green Belts as well as lead improvement projects using specified methodologies such as DMAIC (define, measure, analyze, improve, control), DMADV (define, measure, analyze, design, verify), and DFSS (design for Six Sigma). [I.B.1]

27. d; The Six Sigma role of coach is involved in giving on-the-spot training, advising team leaders, and helping a team prepare for tollgate meetings. [I.B.1]

28. c; A Master Black Belt (MBB) is a Six Sigma role associated with an individual typically assigned full-time to train and mentor Black Belts as well as lead the strategy to ensure that the improvement projects chartered are the right strategic projects for the organization. Master Black Belts are usually the authorizing body to certify Green Belts and Black Belts. [I.B.1]

29. b; Lewin's three-stage change management theory includes the steps: unfreeze, change, and freeze. [I.B.2]

30. c; Successful Six Sigma implementations involve a steering committee with top management representation and support. The implementation is driven from the top down. [I.B.2]

31. a; Effective communication is considered essential for success. Communicating the need for change is the first step in the change management process. [I.B.2]

32. b; A readiness assessment can be used to determine an organization's willingness to change. [I.B.2]

33. c; A force-field analysis lists the project or goal and then brainstorms the forces that can drive or restrain success. [I.B.2]

34. d; Effective communication is considered essential for success. Communicating the need for change is the first step in the change management process. [I.B.2.]

35. c; A highly centralized organizational structure tends to resist any change, and fundamental improvements are difficult to achieve. [I.B.2]

36. a; Suppliers–inputs–process–outputs–customers (SIPOC) can be used to identify stakeholders. [I.B.2]

37. a; Frequent communication of project status throughout the change cycle is the key to maintaining support. [I.B.2]

Section II

Organizational Process Management and Measures

II. Organizational Process Management and Measures (12 Questions)

A. ***Impact on stakeholders.*** Describe the impact six sigma projects can have on customers, suppliers, and other stakeholders. (Understand)

B. ***Benchmarking.*** Define and distinguish between various types of benchmarking, e.g., best practices, competitive, collaborative, and breakthrough. Select measures and performance goals for projects resulting from benchmarking activities. (Apply)

C. ***Business measures***

1. *Performance measures.* Define and describe balanced scorecard, key performance indicators (KPIs), key behavior indicators (KBIs), objectives and key results (OKRs), customer loyalty metrics, and leading and lagging indicators. Explain how to create a line of sight from performance measures to organizational strategies. (Analyze)

2. *Financial measures.* Define and use revenue growth, market share, margin, net present value (NPV), return on investment (ROI), and cost–benefit analysis (CBA). Explain the difference between hard cost measures (from profit-and-loss statements) and soft cost benefits of cost avoidance and reduction. (Apply)

QUESTIONS

1. Any individual or group with an interest in the business is considered a/an:
   a. stakeholder.
   b. shareholder.
   c. regulator.
   d. employee.
2. Delivering the right level of information, to the right stakeholders, at the right time is part of the:
   a. company newsletter.
   b. company annual report.
   c. human resources function.
   d. communications management process.
3. Which of the following forms of communication can be best used to acknowledge a key milestone?
   a. Newsletter
   b. Annual report
   c. Quarterly sales meeting
   d. Announcements
4. A Six Sigma improvement project will address inefficiencies in a painting process. The frontline workers in the paint department are identified as project stakeholders. Which categories of Influence and Importance would likely be assigned to these frontline workers?
   a. High Influence and High Importance
   b. High Influence and Low Importance
   c. Low Influence and High Importance
   d. Low Influence and Low Importance
5. If a project stakeholder is deemed to have High Influence and Low Importance, what behavior approach should the project leader adopt for this stakeholder?
   a. Keep informed to prevent sabotage
   b. Collaborate

c. Protect and defend

d. Do not expend resources

6. A stakeholder who has been designated as low potential for cooperation and high potential to impact a project would be considered:

a. nonsupportive.

b. supportive.

c. a mixed blessing.

d. marginal.

7. A stakeholder whose daily work routine will likely be affected by a project should be:

a. involved in relevant discussion and decisions.

b. handled in a defensive way.

c. informed of changes after the fact.

d. trained as a Black Belt.

8. A current attitude assessment for a stakeholder shows that they are suspicious of the project. The project leader should:

a. keep the stakeholder from working on the team.

b. assign the stakeholder to another department.

c. cancel the project due to resistance.

d. develop an action plan to reduce opposition and increase support.

9. Benchmarking can be used to:

a. apply statistical techniques to control a process.

b. compare the current state to best-in-class processes or products.

c. improve processes using teams of five to seven employees.

d. restructure an entire organization and its processes.

10. Organizations usually use benchmarking to:

a. collect and develop data on industry best practices.

b. perform root cause analysis.

c. drive breakthrough improvement.

d. determine critical-to-quality features.

11. Which of the following actions is the first step a company should take when benchmarking?
    a. Incorporate best practices
    b. Decide what to benchmark
    c. Collect external data
    d. Collect internal data
12. The act of conducting performance or process benchmarking within an organization by comparing similar business units or business processes is known as:
    a. functional benchmarking.
    b. internal benchmarking.
    c. strategic benchmarking.
    d. focused benchmarking.
13. The benchmarking process can be ineffective if it:
    a. has not been fully defined.
    b. is fully mapped.
    c. is not industry specific.
    d. was provided by a consultant.
14. The benchmarking process usually follows which sequence?
    a. Presentation, data collection, statistical analysis, and planning
    b. Presentation, analysis, and implementation
    c. Planning, data collection, analysis, and implementation
    d. Planning, detection factors, statistical analysis, and integration
15. A possible negative effect of benchmarking includes:
    a. breaches of confidentiality.
    b. importing an incompatible culture.
    c. implementing inappropriate practices.
    d. implementing breakthrough improvements.

16. In a balanced scorecard, which perspective addresses shareholder value?
    a. Financial perspective
    b. Learning and growth perspective
    c. Internal business processes perspective
    d. Customer perspective
17. An organization's net promoter score is positively correlated to which metric?
    a. Customer referrals
    b. Customer complaints
    c. Customer abandonment rate
    d. Competitive market share
18. The balanced scorecard incorporates which four perspectives?
    a. Financial, quality, external business relations, and training
    b. Financial, customer, internal business processes, and innovation and learning
    c. Quality, internal business processes, innovation and learning, and competitive advantage
    d. Quality, customer, stakeholders, and innovation and learning
19. Typical metrics used in the internal business processes perspective of the balanced scorecard include:
    a. throughput.
    b. net promoter score.
    c. sustainability.
    d. cash flow.
20. Key process indicators (KPIs) are measured, from the lowest to highest level, as:
    a. operational, tactical, and strategic.
    b. tactical, operational, and strategic.
    c. tactical, financial, and strategic.
    d. operational, tactical, and financial.

21. An example of a tactical KPI for a call center would include:
    a. monthly average hold times.
    b. control chart measurements for call length.
    c. customer satisfaction.
    d. daily call totals.
22. Which perspective provides shareholders with a direct line of sight into the health and well-being of the organization?
    a. Financial perspective
    b. Customer perspective
    c. Internal business processes perspective
    d. Learning and growth perspective
23. Effective KPIs are:
    a. independent of upstream and downstream processes.
    b. complex.
    c. many in number.
    d. standardized.
24. A trucking company strives to improve its on-time delivery. In a staff meeting, it is reported that the on-time delivery rate has increased, and transportation costs have also increased due to trucks leaving the facility half-full. This is an example of:
    a. Six Sigma improvement.
    b. suboptimized systems.
    c. increased failure rate.
    d. unbalanced scorecard.
25. One of the functions of key behavior indicators is to:
    a. track cultural change in an organization.
    b. punish low performers.
    c. dictate social norms.
    d. increase regulatory compliance.

26. Which of the following statements is true about objectives and key results (OKRs)?

    a. OKRs are used to reduce the workforce equitably.

    b. OKRs list objectives and the results required to achieve them.

    c. OKRs incorporate benchmarking metrics.

    d. OKRs cannot be used to track individual achievement.

27. A customer satisfaction survey asks respondents how likely they are to recommend the product to a friend or colleague using a Likert scale in which 1 maps to *not likely at all* and 10 maps to *extremely likely*. Based on the results shown below, what is the value of the net promoter score?

Responses	Count
9–10	23
7–8	12
5–6	5
3–4	4
1–2	1

    a. 28.9%

    b. 51.1%

    c. 55.6%

    d. 77.8%

28. A company monitors the number of hits on its website sales page. This metric is likely a leading indicator of:

    a. future sales.

    b. customer returns.

    c. warranty charges.

    d. customer retention.

29. Which perspective defines an organization's value proposition and measures how effective the organization is in creating value?

    a. Internal business processes perspective

    b. Learning and growth perspective

    c. Financial perspective

    d. Customer perspective

30. Which perspective includes the capabilities and skills of an organization?
    a. Financial perspective
    b. Learning and growth perspective
    c. Internal business process perspective
    d. Customer perspective

31. In five years, $5000 will be available. To the nearest dollar, what is the net present value (NPV) of that money, assuming an annual interest rate of 10%?
    a. $500
    b. $1581
    c. $3105
    d. $5000

32. The net present value (NPV) of the cost to conduct a project is estimated to be $100,000. The NPV of the benefits of the project are estimated at $750,000. What is the benefit-to-cost ratio?
    a. 0.13
    b. 7.5
    c. 100,000
    d. 750,000

33. Costs associated with customer complaints are best characterized as:
    a. appraisal costs.
    b. prevention costs.
    c. internal failure costs.
    d. external failure costs.

34. To the nearest dollar, what is the net present value of a $15,000 cost that will be incurred in three years, assuming a 4.0% interest rate and continuous compounding?
    a. $5000
    b. $13,304
    c. $13,335
    d. $14,416

35. Calculate the discounted return on investment (ROI) for a project with the following cash flows. Assume an interest rate of 5%.

Year	Cost	Benefit
0	$75,000	$150,000
1	$15,000	$45,000
2	$15,000	$45,000

a. $76,741

b. $80,578

c. $80,782

d. $81,000

36. To the nearest dollar, calculate the net present value of a payment of $300,000 that will occur at the end of Year 1, assuming an annual interest rate of 4%, compounded quarterly.

a. $258,621

b. $285,440

c. $288,000

d. $288,294

37. Product A sells for $55.50 and its total cost is $15.00. What is its percentage margin?

a. 27%

b. 37%

c. 73%

d. 270%

38. A two-month improvement yielded $40,000 in benefits after an implementation cost of $25,000. What is the simple ROI for the project?

a. 26%

b. 37.5%

c. 60%

d. 62.5%

SOLUTIONS

1. a; A stakeholder is defined as anyone with an interest in an issue or anyone who can influence or be affected by an action or change. [II.A]

2. d; The communication management process ensures delivering the right level of information to the right stakeholders at the right time. [II.A]

3. d; Announcements (formal or informal) are communications used to acknowledge a key milestone. [II.A]

4. c; The frontline workers for the paint project would likely have low influence in the organization but would be highly important to the project's success. [II.A]

5. a; If a stakeholder has high influence in the organization, but low importance to the project success, they are likely high-level administrators or finance workers. Therefore, a project leader should adopt the "inform to prevent sabotage" behavior approach for these stakeholders. [II.A]

6. a; A stakeholder who has low potential for cooperation and high potential to impact a project would be designated as nonsupportive. [II.A]

7. a; A stakeholder whose daily work routine is likely to be affected by a Six Sigma project should be included in relevant discussion and decisions. [II.A]

8. d; If a stakeholder's current attitude shows that they are suspicious of a project, the team leader should develop an action plan to reduce the stakeholder's opposition and increase their support. [II.A]

9. b; Benchmarking is an improvement process in which a company measures its performance against that of best-in-class companies, determines how those companies achieved their performance levels, and uses the information to improve its own performance. The areas that can be benchmarked include strategies, operations, processes, and procedures. [II.B]

10. c; Benchmarking provides an organization with the opportunity to see what level of process performance is possible and to drive breakthrough improvement over continuous improvement. [II.B]

11. b; When planning to benchmark, an organization must first decide what to benchmark. [II.B]

12. b; Internal benchmarking is the act of conducting performance or process benchmarking within an organization by comparing similar business units or business processes. [II.B]

13. a; Benchmarking requires the use of a strict methodology. It must be fully defined, planned, and funded, or it will likely fail. [II.B]

14. c; The benchmarking process follows the cycle of planning, data collection, analysis, and implementation. [II.B]

15. a; A possible negative effect of benchmarking is that organizations may become exposed to potential breaches of confidentiality. Trade secrets may be involved, and care must be taken to establish a code of conduct for the benchmarking exercise. [II.B]

16. a; The financial perspective provides insight into how well goals, objectives, and strategies are contributing to shareholder value and the bottom line. [II.C.1]

17. a; The net promoter score (NPS) measures the extent to which customers are enthusiastic about the brand and will refer the brand to others. A higher NPS is positively correlated with more customer referrals. [II.C.1]

18. b; The four perspectives addressed in the balanced scorecard include financial, customer, internal business processes, and innovation and learning. [II.C.1]

19. a; Typical metrics for the internal business processes perspective of the balanced scorecard include throughput, cycle time, unit cost, defect rates, and quality level. [II.C.1]

20. a; Key process indictors (KPIs) are measured from the lowest level to the highest level according to operational, tactical, and strategic. [II.C.1]

21. a; Tactical KPIs are used by first-level managers to track progress and make decisions. A tactical KPI for a call center would include monthly average hold times. Control charts and daily call totals would be considered operational, and customer satisfaction would be considered strategic. [II.C.1]

22. a; The financial perspective provides shareholders with a direct line of sight into the health and well-being of the organization. [II.C.1]

23. d; Effective KPIs are aligned to the business's goals; owned in terms of data collection and responsibility to act; predictive; actionable; limited in number; transparent; balanced and linked; standardized; and timely. [II.C.1]

24. b; Concentrating solely on on-time delivery rates is an example of a suboptimized system. When improvements are proposed, it is important to think systemically, so that optimizing one metric will not result in the degradation of another. [II.C.1]

25. a; Key behavior indicators (KBIs) are used to track behavior that can lead to high-level performance. This includes tracking cultural change in an organization. [II.C.1]

26. b; OKRs list objectives for either an organization, a department, or an individual, and then list the results that will be required to achieve the objective. [II.C.1]

27. a; The net promoter score is calculated as the percentage of respondents who are promoters (scores of 9 or 10) minus the percentage of respondents who are detractors (scores from 1–6).

$$NPS = \frac{23-10}{45} = 28.9\%$$

[II.C.1]

28. a: The number of hits on the company's sales page is likely a leading indicator of future sales. [II.C.1]

29. a; The internal business processes perspective defines an organization's value proposition and measures how effective the organization is in creating value. [II.C.1]

30. b; The learning and growth perspective includes the capability and skills of an organization. [II.C.1]

31. c; To calculate net present value (NPV), we note that A = \$5000, r = 0.10, and n = 5. Then

$$NPV = \frac{A}{(1+r)^n}$$
$$= \frac{\$5000}{(1+0.10)^5} = \frac{\$5000}{1.10^5} = \$3105.$$

[II.C.1]

32. b; The benefit-to-cost ratio with an NPV of benefits equal to \$750,000 and NPV of costs equal to \$100,000 is calculated below. [II.C.1]

$$BC = \frac{NPV(Benefit)}{NPV(Cost)}$$
$$= \frac{\$750{,}000}{\$100{,}000} = 7.5$$

33. d; Customer complaints are considered external quality costs. Recall that there are four costs of quality: internal, external, appraisal, and prevention. [II.C.2]

34. c; To calculate net present value (NPV), we note that A = \$15,000, r = 0.04, and n = 3. Then

$$NPV = \frac{A}{(1+r)^n}$$
$$= \frac{\$15{,}000}{(1+0.04)^3} = \frac{\$15{,}000}{1.04^3} = \$13{,}335.$$

[II.C.1]

35. To calculate the discounted return on investment (ROI), we first calculate the net gain for each year:

Year	Cost	Benefit	Net Benefit = Benefit – Cost
0	$75,000	$150,000	$75,000
1	$15,000	$45,000	$30,000
2	$15,000	$45,000	$30,000

The net benefit for Year 0 is $75,000. Year 0 implies that this is present day, so no adjustment is needed for the Year 0. For Years 1 and 2, we will need to bring these figures back to the present. Noting that r = 5%, and t = 0, 1, and 2, we use the formula for discounted ROI:

$$Disc\,ROI = \frac{NPV(Net\,benefit)}{NPV(Cost)} =$$

$$= \frac{\$30,000}{(1+0.05)^3} + \frac{\$30,000}{(1+0.05)^3} = \frac{\$15,000}{1.04^3} = \$13,335.$$

[II.C.2]

36. d; We can calculate the NPV of the payment by noting that A = $300,000, r = 4%, n = 1, and t = 4, which means that compounding occurs four times per year. The quarterly interest rate is then (r/t) = (0.04/4) = 0.01. Then

$$NPV = \frac{A}{\left(1+{}^{r}\!/_{t}\right)^{nt}}$$

$$= \frac{\$300,000}{\left(1+{}^{0.04}\!/_{4}\right)^{(1\times4)}} = \frac{\$300,000}{1.01^4} = \$288,294.$$

[II.C.2]

37. c; We can calculate percentage margin by noting that revenue is equal to $55.50, and the cost is $15.00. Then

$$Percent\,margin = \left(\frac{Revenue - Cost}{Revenue}\right) \times 100\%$$

$$= \left(\frac{\$55.50 - \$15.00}{\$55.50}\right) = 73\%.$$

[II.C.2]

38. c; We note that the project benefit is equal to $40,000 and the cost is $25,000. The project is short-term, so there is no need to adjust the benefit or cost for the time value of money. We then can calculate the simple ROI using the following:

$$Simple\,ROI = \left(\frac{Benefit - Cost}{Cost}\right)$$

$$= \left(\frac{\$40,000 - \$25,000}{\$25,000}\right) = 0.60 = 60\%.$$

[II.C.2]

Section III

Team Management

III. Team Management (15 Questions)

A. ***Team formation***

1. *Team types and constraints.* Define and describe various teams, including virtual, cross-functional, and self-directed. Determine what team type will work best for a given set of constraints, e.g., geography, technology availability, staff schedules, time zones. (Apply)
2. *Team roles and responsibilities.* Define and describe various team roles and responsibilities for leader, facilitator, coach, and individual member. (Understand)
3. *Team member selection criteria.* Describe various factors that influence the selection of team members, including the ability to influence, openness to change, required skills sets, subject matter expertise, and availability. (Apply)
4. *Team success factors.* Identify and describe the elements necessary for successful teams, e.g., management support, clear goals, ground rules, and timelines. (Apply)

B. ***Team facilitation***

1. *Motivational techniques.* Describe and apply techniques to motivate team members. Identify factors that can demotivate team members and describe techniques to overcome them. (Apply)
2. *Team stages of development.* Identify and describe the classic stages of team development: forming, storming, norming, performing, and adjourning. (Apply)
3. *Team communication.* Describe and explain the elements of an effective communication plan, e.g., audience identification, message type, medium, frequency. (Apply)
4. *Team leadership models.* Describe and select appropriate leadership approaches (e.g., direct, coach, support, and delegate) to ensure team success. (Apply)

C. ***Team dynamics***

1. *Group behaviors.* Identify and use various conflict resolution techniques (e.g., coaching, mentoring, and intervention) to overcome negative group dynamics, including dominant and reluctant participants, groupthink, rushing to finish, and digressions. (Evaluate)
2. *Meeting management.* Select and use various meeting management techniques, including using agendas, starting on time, requiring pre-work by attendees, and ensuring that the right people and resources are available. (Apply)
3. *Team decision-making methods.* Define, select, and use various tools (e.g., consensus, nominal group technique, and multi-voting) for decision making. (Apply)

D. ***Team training***

1. *Needs assessment.* Identify the steps involved to implement an effective training curriculum: identify skills gaps, develop learning objectives, prepare a training plan, and develop training materials. (Understand)
2. *Delivery.* Describe various techniques used to deliver effective training, including adult learning theory, soft skills, and modes of learning. (Understand)
3. *Evaluation.* Describe various techniques to evaluate training, including evaluation planning, feedback surveys, pre-training, and post-training testing. (Understand)

QUESTIONS

1. An advantage of a virtual team is its:
   a. reduced technology costs.
   b. asynchronous nature of communication.
   c. emphasis on small teams.
   d. increased flexibility.
2. A company has formed a team that has members with diverse roles and skill sets. This type of team is best defined as:
   a. unfocused.
   b. process improvement.
   c. cross-functional.
   d. self-directed.
3. Which of the following individuals does not generally attend regular Six Sigma project team meetings?
   a. Facilitator
   b. Team leader
   c. Scribe
   d. Champion
4. The team coach for an improvement project is usually responsible for:
   a. maintaining and publishing meeting minutes.
   b. summarizing progress using visual aids.
   c. notifying the team leader or facilitator of the time remaining on each agenda item.
   d. working with the team leader and facilitator to move the team toward the objective.
5. The person providing assistance when necessary to individual team members to ensure that action items are completed is the:
   a. coach.
   b. champion.
   c. leader.
   d. facilitator.

6. The individual responsible for organizing, hosting, and running the team meeting is the:

   a. champion.

   b. sponsor.

   c. mentor.

   d. leader.

7. An effective Six Sigma project team member:

   a. always agrees with the group.

   b. avoids constructive debate.

   c. completes action assignments on time.

   d. engages in concurrent discussions.

8. When choosing team members, it is important to select those who:

   a. are open to change.

   b. have the ability to dominate discussions.

   c. agree with the group.

   d. make decisions using an authoritative approach.

9. Process improvement teams should have training on which of the following topics?

   a. Process dynamics

   b. Team building

   c. Project management

   d. Program management

10. Which of the following items can best contribute to team success?

   a. A sponsor with personal stake in the project

   b. A Yellow Belt in attendance at all meetings

   c. Required use of designed experiments

   d. Advanced statistical training for all team members

11. Which of the following factors is considered a demotivating factor for team members?
    a. Monetary rewards
    b. Changes in project scope
    c. Too many resources
    d. Public expressions of appreciation
12. Lencioni describes five dysfunctions of a team. Which dysfunction is considered the primary inhibitor of a successful team?
    a. Absence of trust
    b. Fear of conflict
    c. Avoidance of accountability
    d. Lack of commitment
13. According to Lencioni, before lack of commitment can be addressed, which dysfunctions must be overcome?
    a. Inattention to team results and lack of success
    b. Fear of conflict and absence of trust
    c. Avoidance of accountability and absence of trust
    d. Lack of success and fear of conflict
14. Team members working on a process improvement project struggle to understand the team's goal and its meaning for them individually. The team is likely in which stage of team development?
    a. Forming
    b. Performing
    c. Storming
    d. Norming
15. After the first four team growth stages, a team will pass through which two final stages?
    a. Forming and norming
    b. Rewarding and recognition
    c. Performing and adjourning
    d. Adjourning and recognition

16. The team phase in which team members begin to understand the need to operate as a team rather than as a group of individuals is referred to as:
    a. storming.
    b. performing.
    c. forming.
    d. norming.
17. A communication plan includes which of the following elements?
    a. A problem statement and Gantt chart
    b. How messages are to be delivered and how often
    c. Team members' influence and importance ratings
    d. The performance gap and measurable item
18. What is one of the most frequently cited causes for team failure?
    a. The use of too many charts
    b. Inadequate communication
    c. The use of SMART goals
    d. Lack of advanced statistical knowledge
19. Communication outside the team:
    a. is not necessary.
    b. should be detailed and lengthy.
    c. should occur only after project completion.
    d. can be delivered via a status report.
20. Which type of leader provides clear instructions to team members?
    a. Supportive
    b. Coaching
    c. Empowering
    d. Directional
21. A leader who empowers process improvement team members to function independently is using which leadership style?
    a. Delegating
    b. Supporting

c. Directing

d. Consulting

22. What type of leadership is best suited for team members with average competence and moderate commitment?

    a. Delegating

    b. Facilitating

    c. Directing

    d. Supporting

23. Which leadership characteristics are best used for team members with low competence but high commitment?

    a. Low directive and low supportive

    b. High directive and low supportive

    c. Low directive and high supportive

    d. High directive and high supportive

24. Which of the following statements concerning the role of a coach or mentor is correct?

    a. Coaches usually work with individuals to develop skills and abilities.

    b. Mentors usually work with teams to develop skills and abilities.

    c. Mentors help projects stay on track by providing resources for completion of team member assignments.

    d. Coaches provide guidance and direction on how to navigate organizational barriers.

25. In a team meeting, one member continually has private conversations with the person next to them. This type of disruptive team member is known as the:

    a. whisperer.

    b. naysayer.

    c. dropout.

    d. team clown.

26. A team member who agrees with whomever they are currently speaking with and views decisions as a way to get ahead is known as the:

    a. whisperer.

    b. politician.

c. storyteller.

d. naysayer.

27. A team leader seeks out dissenting views in a team meeting. The leader is likely trying to get the team to avoid which behavior?

    a. Participating in groupthink

    b. Airing old disputes

    c. Accepting opinions without data

    d. Arguing over who gets credit

28. A team leader goes around the conference table during a brainstorming session, asking each person for one comment. The leader is likely trying to avoid:

    a. scope creep.

    b. running out of time.

    c. one person from dominating the discussion.

    d. accepting opinions without data.

29. A team gets off on a tangent when discussing an improvement option. The team facilitator can bring the team's focus back by using which technique?

    a. Boomerang

    b. Redirect

    c. Idea floating

    d. Direct probe

30. A team member is uncooperative and unassertive. They are exhibiting which type of conflict management behavior?

    a. Avoiding

    b. Accommodating

    c. Compromising

    d. Collaborating

31. What sets the mentor role apart from the coaching role?

    a. A mentor helps individuals choose the proper tools to use in Six Sigma projects.

    b. A mentor focuses on the individual from a career perspective.

c. A mentor is a Master Black Belt.

d. A mentor concentrates on project-specific issues.

32. A team is having trouble generating ideas for improvement. The team facilitator can spark the team's creativity by using which technique?

    a. Lasso

    b. Redirect

    c. Idea floating

    d. Direct probe

33. Which of the following behaviors should a facilitator avoid?

    a. Being diligent about time keeping

    b. Permitting digressions

    c. Acting in an unbiased manner

    d. Speaking in a steady, non-emotional voice

34. When a team rushes to meet its milestones without thorough analysis, it can run the risk of:

    a. optimizing the process.

    b. missing a root cause.

    c. finishing under budget.

    d. ignoring ground rules.

35. Rank the following methods for defusing disruptive behavior, from the lowest to the highest on the escalation scale.

    1. Stand up and walk halfway

    2. Pull aside during a break

    3. Walk by while making eye contact

    4. Call out behavior in front of the team

    a. 1, 2, 3, 4

    b. 2, 3, 1, 4

    c. 3, 1, 2, 4

    d. 1, 3, 2, 4

36. Requiring pre-work, creating team norms, and conducting end-of-meeting process checks are all methods that can be used to:

    a. make team decisions.

    b. keep a team productive.

    c. slow team progress.

    d. gather voice of the customer data.

37. Inviting a subject matter expert to attend a team meeting when needed is a way to:

    a. upset team norms.

    b. move the team to the performing stage.

    c. ensure the right people and resources are available.

    d. postpone team decisions.

38. A project team has discussed several possible solutions based on data provided by the process owner. The team has reached agreement on the course of action. This agreement type is referred to as:

    a. consensus.

    b. decisive.

    c. one-sided.

    d. dysfunctional.

39. A team generates ideas by writing them down without discussion. Afterward, the ideas are ranked. This technique is known as:

    a. brainstorming.

    b. affinity analysis.

    c. nominal group technique.

    d. consensus building.

40. Brainstorming is most associated with

    a. generating ideas.

    b. eliminating ideas.

    c. ranking ideas.

    d. reducing ideas.

41. The nominal group technique is a decision-making method that
    a. creates a Pareto chart of top project ideas.
    b. incorporates a prioritization matrix.
    c. arranges the group ideas in categories.
    d. shows relationships between input and outputs.
42. The terms *training* and *education* are often used interchangeably. However, training is best described as:
    a. skill-based instruction.
    b. knowledge-based instruction.
    c. increasing an understanding of concepts.
    d. preparation for future opportunities.
43. The training development process begins with which step?
    a. Performing a training needs analysis
    b. Ensuring that the budget has been approved
    c. Scheduling the training activities
    d. Developing the training materials
44. When using the ADDIE training development model, which of the following actions occur during the design phase?
    a. A project charter and problem statement are created.
    b. A training blueprint is created and group exercises are planned.
    c. Designers create storyboards for the training.
    d. Designers identify learners and their learning constraints.
45. The ADDIE training development model includes which five phases?
    a. Analysis, design, development, implementation, and evaluation
    b. Analysis, development, deployment, improvement, and evaluation
    c. Analysis, definition, deployment, improvement, and evaluation
    d. Analysis, description, development, implementation, and evaluation

46. Which item should be included in a training needs assessment?
    a. Types of classroom materials required
    b. Final test questions
    c. Detailed training plans
    d. The business case for the training need
47. One use of a training matrix includes:
    a. selecting team members according to their skill sets.
    b. punishing low performers.
    c. creating competition between departments.
    d. showing management that employees are unskilled.
48. According to Ebbinghaus, what is the learning retention rate for a newly acquired, but unused skill after two days?
    a. 21%
    b. 28%
    c. 34%
    d. 50%
49. Which statement is true about online training?
    a. The course design should mirror the in-person version.
    b. Instructors can incorporate group exercises.
    c. Online training must be asynchronous.
    d. Online training does not work well for remote workers.
50. Dr. Deming's red-bead experiment is an example of what type of training exercise?
    a. Asynchronous
    b. Ineffective
    c. Role-playing
    d. Case study

51. One method for measuring training effectiveness includes:

    a. on-the-job observation.

    b. Kano surveys.

    c. Pareto analysis.

    d. salary surveys.

52. To evaluate the effectiveness of a training program, it is important to determine whether the training impacted business results and whether:

    a. technology was used effectively.

    b. the trainer's delivery was entertaining.

    c. the training changed the learner's behavior.

    d. the post-test scores are lower than the pre-test scores.

SOLUTIONS

1. d; Virtual teams increase flexibility. These types of teams can include members from different geographical locations and areas of the organization without incurring travel expenses. [III.A.1]
2. c; Cross-functional teams involve representation from various groups likely to be impacted by the changes. [III.A.1]
3. d; Team champions do not generally attend regular team meetings. [III.A.2]
4. d; Coaches work with the team leader and facilitator to move the team toward the objective and to help provide resources for the completion of team member assignments. Additionally, they evaluate the team's progress. [III.A.2]
5. a; A coach helps provide resources for completion of team member assignments. [III.A.2]
6. d; The team leader is responsible for organizing, hosting, and running the team meeting. [III.A.2]
7. c; Effective team members are characterized by actively participating in team meetings, communicating ideas and expertise, openly listening, and completing action assignments as scheduled. [III.A.3]
8. a; Effective team members should be open to change. In addition, good choices for team members would be those who can influence others, have subject-matter expertise, have availability, and who are team-oriented.
9. b; A process improvement team should have training on team building. Without an understanding of how a team works and the individual behaviors that advance team progress, the team will often get caught in personality concerns and turf wars. [III.A.4]
10. a; A sponsor with a vested interest in the project can help remove barriers and roadblocks for the team, thereby setting the team up for success. [III.A.4]
11. b; Changes in project scope can become a demotivating factor for a team. [III.B.1]
12. a; According to Lencioni, absence of trust is the primary dysfunction of a team and an inhibitor of team success. The bedrock of a well-functioning team much be trust. Team members must feel comfortable sharing their opinions and making mistakes. In the absence of trust, no other dysfunctions can be overcome. [III.B.1]
13. b; According to Lencioni, fear of conflict, as well as absence of trust, must be addressed before lack of commitment can be addressed. [III.B.1]

14. a; Team forming is the stage in which members struggle to understand the team goal and its meaning for them individually. [III.B.2]

15. d; After the first four team phases, Adjourning was added as a final stage. This stage acknowledges the "life cycle" of a team, in which there is a distinct end point for the team. Then the Recognition phase was added to ensure that the team is properly recognized for its accomplishments. [III.B.2]

16. d; During the norming stage, members begin to understand the need to operate as a team rather than as a group of individuals. [III.B.2]

17. b; A communication plan should include who, what, how and how often messages are to be delivered to an individual or group. [III.B.3]

18. b; Inadequate communication is one of the most frequently cited cases for team failure. [III.B.3]

19. d; Communication outside the team should occur regularly. It can take the form of a bulletized status report, or a presentation. Types of items covered include accomplishments, problems, issues or concerns, actions being taken and 30-day plans. [III.B.3]

20. d; Directional leadership is characterized by high directive and low supportive behavior. This leadership style is best suited for dealing with individuals who possess low competence and high commitment. Such individuals are often described as *enthusiastic beginners.* They are new to a task and motivated about it. [III.B.4]

21. a; The delegating leadership style is characterized by low directive and low supportive behavior. It is suited for individuals who possess high competence and high commitment. Such individuals are often described as *self-reliant achievers.* They are proficient in their task and are highly motivated about it. [III.B.4]

22. d; The supporting leadership style is characterized by low directive and high supportive behavior. It is suited for individuals who possess mid-competence and moderate commitment. Such individuals are often described as *capable, but cautious, performers.* They know their task but are not motivated about it. [III.B.4]

23. b; For team members who have low competence but high commitment, a directing leadership style is best. This style is characterized by low supportive but high directive leadership. [III.B.4]

24. a; Coaches work with individuals to develop skills and abilities. They also work with teams to help projects stay on track by providing resources for completion of team member assignments. Mentors work with individuals, not teams, and help mentees navigate organizational barriers and provide guidance and direction for an individual's career. [III.C.1]

25. a; The whisperer has private conservations with their neighbors during meetings. Others may feel that the whisperer is speaking negatively about other team members. [III.C.1]

26. b; The politician sees every decision as a way to get ahead and may agree with whomever they are currently speaking with to curry favor. [III.C.1]

27. a; The leader is seeking out dissenting views to avoid groupthink. [III.C.1]

28. c; To avoid one person dominating the conversation, the team leader can use a "round robin" brainstorming approach, in which each person is asked for one comment in turn. [III.C.1]

29. b; A team leader or facilitator can get the team back on track by using a redirect technique. [III.C.1]

30. a; In the avoiding approach to conflict management, the participant hopes the problem will go away. It is used when team members are uncooperative and unassertive. [III.C.1]

31. b; A mentor focuses on an individual from a career perspective; a coach focuses on the individual concerning issues involving Lean Six Sigma. [III.C.1]

32. c; A facilitator can use the idea-floating method to spark the team's creativity.

33. b; A facilitator should not permit the team to get off track and should discourage and stop digressions. [III.C.1]

34. b; When a team rushes to complete milestones, it runs the risk of missing root causes because it is not thoroughly analyzing information. [III.C.1]

35. d; In order from lowest to highest on the escalation scale, the actions are (1) stand up and walk halfway, (2) walk by while making eye contact, (3) pull aside during a break, and (4) call out behavior in front of the team. [III.C.1]

36. b; Requiring pre-work, creating team norms, and conducting end-of-meeting process checks are all methods that can be used to keep a team productive. [III.C.2]

37. c; Inviting a subject matter expert to attend a team meeting when needed is a way to ensure the right people and resources are available. [III.C.2]

38. a; The consensus agreement type creates a situation or outcome that all the participants can live with. If no such situation or outcome can be found, discussion continues. [III.C.3]

39. c; The nominal group technique (NGT) is used to prioritize a list of items. The technique begins with all team members writing their ideas down on a slip of paper without any discussion. Later, the ideas are ranked. [III.C.3]

40. a; Brainstorming is a group process used to generate ideas in a nonjudgmental way. The purpose of brainstorming is to generate a large number of ideas about an issue. [III.C.3]

41. b; The nominal group technique (NGT) incorporates a prioritization matrix to allow the team to decide among options. [III.C.3]

42. a; Education focuses on broadening an individual's knowledge base and expands thinking processes. Training is considered a subset of education that focuses on increasing proficiency in a skill. [III.D.1]

43. a; The training development process begins with performing a training needs analysis. This needs analysis must be conducted before any training plans can be made. [III.D.1]

44. b; The ADDIE training development model has five phases: analysis, design, development, implementation, and evaluation. During the design phase, a blueprint of the training is created, and learning objectives are identified. In addition, the designers plan for the course content and individual and group exercises [III.D.1]

45. a; The ADDIE training development model consists of five phases: analysis, design, development, implementation, and evaluation. [III.D.1]

46. d; A training needs assessment should provide a business case for the training need. [III.D.1]

47. a; A training matrix can be used to select team members according to their skill levels. It is also used to spot inadequacies in skill levels and create training plans for individual employees. [III.D.1]

48. b; According to Ebbinghaus, there is only 28% retention of a new skill that is learned if it is not applied after two days. [III.D.2]

49. b; Advances in online training platforms make it possible for instructors to incorporate group exercises into their online training. [III.D.2]

50. c; Dr. Deming's red-bead experiment is a type of role-playing exercise. [III.D.2]

51. a; On-the-job observation is one way to assess whether a learner's behavior has changed due to training. [III.D.3]

52. c; Effective training should result in a change in the learner's behavior. [III.D.3]

Section IV

Define

IV. Define (20 Questions)

A. *Voice of the customer*

1. *Customer identification.* Identify and segment customers and show how a project will impact both internal and external customers. (Apply)
2. *Customer data collection.* Identify and select appropriate data collection methods (e.g., surveys, focus groups, interviews, and observations) to gather voice of the customer data. Ensure the data collection methods used are reviewed for validity and reliability. (Analyze)
3. *Customer requirements.* Define, select, and apply appropriate tools to determine customer needs and requirements, including critical-to-X (CTX when "X" can be quality, cost, safety, etc.), CTQ tree, quality function deployment (QFD), supplier, input, process, output, customer (SIPOC), and Kano model. (Analyze)

B. ***Business case and project charter***

1. *Business case.* Describe business case justification used to support projects. (Understand)
2. *Problem statement.* Develop a project problem statement and evaluate it in relation to baseline performance and improvement goals. (Evaluate)
3. *Project scope.* Develop and review project boundaries to ensure that the project has value to the customer. (Analyze)
4. *Goals and objectives.* Identify SMART (specific, measurable, actionable, relevant, and time-bound) goals and objectives on the basis of the project's problem statement and scope. (Analyze)
5. *Project performance measurements.* Identify and evaluate performance measurements (e.g., cost, revenue, delivery, schedule, and customer satisfaction) that connect critical elements of the process to key outputs. (Analyze)
6. *Project charter review.* Explain the importance of having periodic project charter reviews with stakeholders. (Understand)

C. ***Project management (PM) tools.*** Identify and use the following PM tools to track projects and document their progress. (Evaluate)

1. Gantt charts
2. Tollgate reviews
3. Work breakdown structure (WBS)
4. RACI Model (responsible, accountable, consulted, and informed)

D. ***Analytical tools.*** Identify and use the following analytical tools throughout the DMAIC cycle. (Apply)

1. Affinity diagrams
2. Tree diagrams
3. Matrix diagrams
4. Prioritization matrices
5. Activity network diagrams
6. Process decision program chart (PDPC)
7. Interrelationship digraph (ID)

QUESTIONS

1. What is the voice of the customer?
   a. The response heard from market promotions
   b. An expression of customer wants
   c. The communication method used for mass marketing
   d. A measure of the capability of a process
2. A hospital IT department is implementing a new electronic medical records system that will make it easier for nurses to complete their reports. The nurses can be thought of as the IT department's:
   a. champions.
   b. facilitators.
   c. internal customers.
   d. external customers.
3. A data collection instrument is determined to be biased. It is likely to be:
   a. reliable but not valid.
   b. reliable and valid.
   c. unreliable but valid.
   d. unreliable and not valid.
4. Interviews, suggestion boxes, focus groups, and surveys are all examples of customer:
   a. listening posts.
   b. specifications.
   c. expectations.
   d. feedback loops.
5. Which of the following sources of customer data is considered external?
   a. Complaints and transaction data
   b. Market research
   c. Advertising media
   d. Warranty usage

6. Mystery shoppers, focus groups, and comment cards are all examples of:
   a. CTQs
   b. Listening posts
   c. Marketing tools
   d. Transaction data
7. Which of the following tools captures requirements for inputs and outputs from a process?
   a. Quality function deployment
   b. Kano model
   c. Process decision program chart
   d. SIPOC diagram
8. Specifications most accurately represent:
   a. natural process limits.
   b. critical-to-quality attributes.
   c. Taguchi's loss function.
   d. the voice of the customer.
9. The house of quality is used to:
   a. organize a company into departments focusing on quality.
   b. structure the engineering team.
   c. translate customer needs into design specifications.
   d. calculate specifications for desired cost reductions.
10. In the Kano model, expected requirements are those that will cause:
    a. low customer dissatisfaction if not met, but high customer satisfaction if met.
    b. high customer dissatisfaction if not met, and high customer satisfaction if met.
    c. high customer dissatisfaction if not met, but neutral customer satisfaction if met.
    d. low customer dissatisfaction if not met, and neutral customer satisfaction if met.

11. A critical-to-quality flow-down:
    a. shows the steps of a process from order to delivery.
    b. shows how customer satisfaction goals can be broken into measurable goals.
    c. gathers CTQs to arrive at a strategic goal.
    d. uses Kano analysis to identify product features.
12. In the house of quality, customer requirements are generally written:
    a. along the far-left side.
    b. on the "roof."
    c. in the interior of the house.
    d. on the bottom of the matrix.
13. In a house of quality, the top portion, or roof, shows the relationship between:
    a. technical requirements.
    b. customer requirements.
    c. technical and customer requirements.
    d. competitors and customer requirements.
14. List, from first to last, the ways in which quality function deployment (QFD) can be used for product planning.
    a. Product, process, and part planning
    b. Product, part, and process planning
    c. Process, product, and part planning
    d. Part, product, and process planning
15. An advantage of a SIPOC diagram is that it:
    a. is more detailed than a value stream map.
    b. can provide quick insights into inputs and outputs.
    c. can be used to track waste.
    d. can be used to determine root causes of a problem.

16. A Kano analysis shows that battery life is a one-dimensional requirement. This means that:

    a. the customer will be delighted with the product.

    b. the longer the battery life, the happier the customer will be.

    c. the customer is indifferent to the length of battery life.

    d. increasing battery life will not improve customer satisfaction.

17. Consistently meeting must-be requirements has what effect on customer satisfaction?

    a. Customer satisfaction will increase.

    b. Customer satisfaction will decrease.

    c. Customer satisfaction will remain steady.

    d. Customers will express delight.

Use the house of quality diagram below to answer questions 18 and 19.

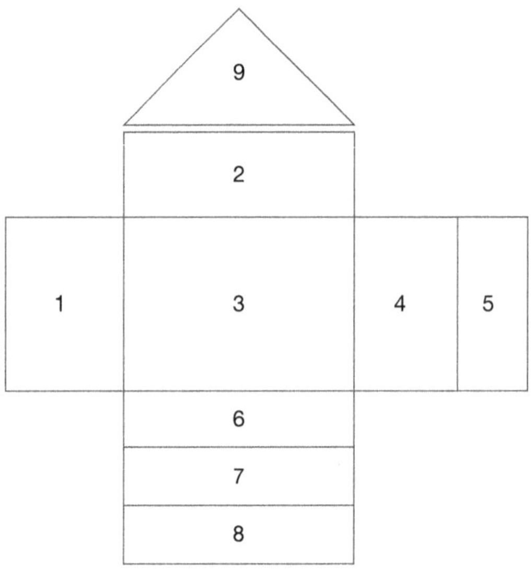

18. The target values for each technical requirement appear in which area of the house of quality?

    a. Area 2

    b. Area 4

    c. Area 7

    d. Area 9

19. Customer requirements are listed in which area of the house of quality?
    a. Area 1
    b. Area 2
    c. Area 4
    d. Area 9
20. The term *CTD* refers to which customer focus?
    a. Critical to defects
    b. Critical to define
    c. Critical to delivery
    d. Critical to design
21. The business results section of a project charter:
    a. includes an estimate of project cost savings.
    b. describes the goals of the project.
    c. lists resources needed for the project.
    d. presents the current and projected future states for project metrics.
22. A well-crafted problem statement will:
    a. include the history of the issue.
    b. use qualitative measures.
    c. present the issue clearly and concisely.
    d. include potential root causes of the issue.
23. A problem statement should contain which elements?
    a. Measurable item, goals, and potential solutions
    b. Business impact, estimated project duration, and goal metrics
    c. Location of the problem, performance gap, and root causes
    d. Measurable item, performance gap, and business impact

24. Evaluate the adequacy of the following problem statement:

    Customer complaints for Product X from Plant 101 have increased by 4% in the last six months.

    a. The problem statement meets all requirements.

    b. The problem statement is missing the root cause.

    c. The problem statement is missing the business impact.

    d. The problem statement is missing the proposed solution.

25. One of the purposes of the project scope is to ensure that the:

    a. root cause is identified.

    b. constraints of the project are clear.

    c. costs of the project are identified.

    d. benefits of the project are clear.

26. Which project scoping method emphasizes the relationship between process inputs and outputs?

    a. Dimensional method

    b. Functional method

    c. SMART method

    d. PERT method

27. The elements of a SMART goal include specific, measurable, and:

    a. achievable, relevant, and time-bound.

    b. action-oriented, relevant, and time-sensitive.

    c. achievable, root cause, and time-bound.

    d. aggressive, reliable, and time-bound.

28. A Six Sigma team is pursuing a project to increase throughput on an injection molding line. The throughput was formerly 600 pieces/hour but has dropped to 500 pieces per hour. Evaluate the following goal statement in terms of a SMART goal:

    Increase throughput to 600/pieces per hour.

    a. The goal statement meets all SMART criteria.

    b. The goal statement is missing relevant.

c. The goal statement is missing time-bound.

d. The goal statement is missing measurable.

29. A Six Sigma project is chartered to increase on-time deliveries. The team has decided that it will track two metrics: on-time deliveries as well as shipping cost. Shipping cost has likely been added to the project metrics to:

 a. appease the finance department.

 b. set the team up for failure.

 c. avoid suboptimizing the process.

 d. satisfy the project champion.

30. An increase in project scope can result in:

 a. an increase in quality and a decrease in cost.

 b. a decrease in duration and quality.

 c. an increase in duration and cost.

 d. a decrease in cost and quality.

31. The project management tool that allows the team to see planned and completed tasks on a timeline is the:

 a. Gantt chart.

 b. affinity diagram.

 c. tree diagram.

 d. interrelationship diagram.

32. In a Gantt chart, task C cannot be started until tasks A and B are completed. Tasks A and B would be considered:

 a. dependencies.

 b. critical tasks.

 c. primary tasks.

 d. secondary tasks.

33. To ensure that all the activities for a particular stage in a process are completed before moving to the next stage, _____________ are needed.

 a. tollgate reviews

 b. Gantt charts

c. interrelationship diagrams

d. work breakdown structures

34. A typical question a champion may ask in a tollgate review for the analyze phase may be:

 a. Will we run a pilot?

 b. What are the critical Xs?

 c. Who are the stakeholders?

 d. What did the FMEA show?

35. What is the purpose of a work breakdown structure?

 a. To locate which machines need repair

 b. To document where the team in failing

 c. To subdivide larger tasks into smaller tasks

 d. To find the critical path of the project plan

36. A tool used to show the activities, sub-activities, costs, time, and materials necessary to complete a goal is the:

 a. tree diagram.

 b. interrelationship diagram.

 c. Gantt chart.

 d. work breakdown structure.

37. In which role in a RACI model must a decision be discussed with the individual before a final decision is made?

 a. Responsible

 b. Accountable

 c. Consulted

 d. Informed

38. In which role in the RACI model must the individual be cognizant about a decision because they are affected?

 a. Responsible

 b. Accountable

 c. Consulted

 d. Informed

39. In an affinity diagram, the themes into which the ideas are grouped:

    a. are suggested from the ideas themselves.

    b. are predetermined.

    c. are generated in a separate brainstorming session on the categories.

    d. are determined by the organizer.

40. Which tool should be used when the issue at hand is ill-defined or difficult to grasp, and group consensus is needed?

    a. Activity network diagram

    b. Process decision program chart

    c. Affinity diagram

    d. Interrelationship digraph

41. Which of the following tools can be used to organize the ideas developed during a brainstorming session into larger categories?

    a. Interrelationship diagram

    b. Gantt chart

    c. Affinity diagram

    d. Work breakdown structure

42. An organization chart is an example of a/an:

    a. interrelationship diagram.

    b. Gantt chart.

    c. scatter plot.

    d. tree diagram.

43. Quality function deployment is an example of a more complicated:

    a. affinity diagram.

    b. Gantt chart.

    c. scatter plot.

    d. matrix diagram.

44. A tool that compares two or more sets of data to discover existing relationships is the:

    a. matrix diagram.

    b. affinity diagram.

    c. Pareto chart.

    d. histogram.

Use the following information to answer questions 45 and 46.

A manufacturer compares manufacturing sites, products, customers, and shipping companies in terms of volume using a matrix diagram, shown below.

	●	Division 1	●	○
●		Division 2		●
Shipper 1	Shipper 2		Part A	Part B
	●	Customer 1	●	
●	○	Customer 2	○	●

● Large volume
○ Small volume

45. What type of matrix is represented here?

    a. L-shaped matrix

    b. T-shaped matrix

    c. X-shaped matrix

    d. Y-shaped matrix

46. Based on the matrix diagram, what can be said about Part B?

    a. Part B is only produced by Division 2.

    b. Division 1 makes the highest volume of Part B.

    c. Customer 2 buys a large volume of Part B.

    d. Division 2 ships Part B using Shipper 2.

47. Which type of matrix can be used to compare pairwise relationships between three variables?

    a. Roof-shaped matrix

    b. L-matrix

    c. T-shaped matrix

    d. X matrix

48. A prioritization matrix was created by a team to choose which software product to purchase. Each criterion was ranked from 1 = best to 4 = worst. Based on the completed matrix, which product should the team consider first?

	Ease of Use	Low Cost	Time to Install
Product	**50%**	**25%**	**25%**
A	4	3	1
B	3	1	2
C	1	2	4
D	2	4	3

a. Product A

b. Product B

c. Product C

d. Product D

Use the precedence diagram below to answer questions 49–51.

Activity	Predecessor	Time
A		5
B		10
C	A	8
D	B	2
E		15
F	C, D, E	6

49. Identify the activities on the critical path.

a. ACF

b. BDE

c. EF

d. CE

50. What is the early start time for Activity C?

a. 5

b. 8

c. 10

d. 13

51. Activity E is delayed by five days; what effect will this have on the overall project timeline?
    a. There is no effect on the finish date.
    b. The other activities can be completed early to maintain the schedule.
    c. Activity C will not be able to start on time.
    d. The project will finish at least five days late.

52. In an activity network diagram, a "late finish" value for an activity represents:
    a. the expected overrun in the project timeline.
    b. the latest that activity can finish without delaying the project.
    c. the slack time for the activity.
    d. the critical path.

53. The slack time for all activities in the critical path is equal to:
    a. the project completion time.
    b. zero.
    c. the longest activity in the path.
    d. the late start time of the first activity.

54. A tool that lists ways a project activity can go wrong and proposed countermeasures is called a/an:
    a. FMEA.
    b. interrelationship digraph.
    c. affinity diagram.
    d. process decision program chart.

55. A process decision program chart begins as a/an:
    a. FMEA.
    b. tree diagram.
    c. fault tree analysis.
    d. cause and effect diagram.

56. The nodes on an interrelation digraph with the most outgoing arrows are considered:

a. drivers.

b. outcomes.

c. contingencies.

d. correlated.

SOLUTIONS

1. b; The voice of the customer is an expression of what the customer wants. [IV.A.1]
2. c; The nurses are the IT department's internal customers since both parties are in the same organization. [IV.A.1]
3. a; A biased data collection instrument is said to be reliable but not valid. [IV.A.2]
4. a; Interviews, suggestion boxes, focus groups, and surveys are all examples of customer data collection mechanisms, or listening posts. [IV.A.2]
5. c; Advertising media, such as television, social media, or radio, are considered external sources of customer data. [IV.A.2]
6. b; Mystery shoppers, focus groups, and comment cards are all types of listening posts from which an organization can collect customer data. [IV.A.2]
7. d; A SIPOC diagram has columns for suppliers, inputs, process steps, outputs, and customers. The I and O columns list the requirements for inputs and outputs of a process, respectively. [IV.A.3]
8. d; Specifications most accurately reflect the customer requirements, or the voice of the customer. [IV.A.3]
9. c; The house of quality matrix used in quality function deployment (QFD) helps illustrate the linkage between the voice of the customer and the resulting technical requirements. Therefore, it is used to translate customer needs into design specifications. [IV.A.3]
10. c; In Kano analysis, expected requirements are known as must-be requirements. Not meeting these requirements will result in high customer dissatisfaction. However, meeting minimum requirements will neither please nor displease the customer. [IV.A.3]
11. b; A critical-to-quality flow-down starts with the high-level strategic goal of customer satisfaction. It then helps the team determine how the larger goal flows down into measurable goals. [IV.A.3]
12. a; The leftmost side of the house of quality lists customer requirements. [IV.A.3]
13. a: The "roof" of the house of quality shows the relationship between pairs of technical requirements using symbols. [IV.A.3]
14. b; Quality function deployment is first used for product planning, then part planning, and finally, process planning. [IV.A.3]
15. b; A SIPOC diagram gives a high-level view of the process and can provide quick insights into inputs and outputs. [IV.A.3]

16. b; A one-dimensional requirement is also known as a satisfier. As the product or service delivers more for this type of requirement, the customer's satisfaction will increase. Delivering more battery life at the same cost would increase customer satisfaction. [IV.A.3]

17. c; Consistently meeting minimum or must-be requirements will cause satisfaction to remain steady. Customer satisfaction will decrease if the must-be requirement is not met, but it will remain neutral if it is met. [IV.A.3]

18. c; The target values for each technical requirement appear in Area 7 of the house of quality. The technical requirements themselves are listed in Area 2. [IV.A.3]

19. a; Customer requirements are listed along the left side of the house of quality, in Area 1. [IV.A.3]

20. c; The term *CTD* refers to critical-to-delivery. Customers who are sensitive to delivery times are identified, and requirements are created to satisfy these customers' delivery needs. [IV.A.3]

21. a; The business results section of a project charter includes an estimate of project cost savings. These savings should be vetted by the finance department. [IV.B.1]

22. c; A well-crafted problem statement presents the issue clearly and concisely. [IV.B.2]

23. d; The problem statement should relay where the problem is occurring, how long it has existed, the measurable item, the performance gap, and the business impact. [IV.B.2]

24. c; The problem statement should include where the problem exists, how long it has existed, the measurable item, the performance gap, and the business impact. The statement given is missing the business impact. [IV.B.2]

25. b; The project scope spells out the constraints of the project, such as budget, time, authority, and resources. [IV.B.3]

26. b; The functional method for characterizing a project scope emphasizes the relationship between input and output variables. This is also referred to as $Y = f(X)$. [IV.B.3]

27. a; SMART goals are specific, measurable, relevant (or realistic), and time-bound. [IV.B.4]

28. c; SMART goals are specific, measurable, relevant (or realistic), and time-bound. The statement *Increase throughput to 600/pieces per hour* is missing the time component. By when will the team achieve its goal? [IV.B.4]

29. c; The team must track both on-time deliveries and shipping costs so that the overall delivery process is not suboptimized. [IV.B.4]

30. c; Increasing project scope can lead to an increase in the project's duration as well as increased costs. [IV.B.5]

31. a; A Gantt chart shows planned and completed tasks on a timeline. [IV.C.1]

32. a; If task C cannot begin before tasks A and B are completed, then A and B are considered dependencies, or predecessor tasks. [IV.C.1]

33. a; Tollgate reviews are formal meetings held with the project sponsor after each of the DMAIC steps. During the meeting, the sponsor ensures that all the activities for a phase are completed before the team moves on to the next phase. [IV.C.2]

34. b; The purpose of the analyze tollgate, among other things, is to present the critical Xs of the process and establish root causes. [IV.C.2]

35. c; In a work breakdown structure, large tasks are systematically broken down into smaller and smaller tasks. These small tasks can then be assigned to individuals. [IV.C.3]

36. d; A work breakdown structure starts with the larger goal and then systematically breaks each activity into smaller and smaller subactivities or tasks. At the final stage, costs, time, and material information are also collected. [IV.C.3]

37. c; The RACI model is a matrix that lists stakeholders and tasks, and then indicates who is responsible, accountable, consulted, and informed for each task. A stakeholder in the "consulted" role must be asked about a decision before a final decision is made. [IV.C.4]

38. d; The RACI model is a matrix that lists stakeholders and tasks, and then indicates who is responsible, accountable, consulted, and informed for each task. A stakeholder in the "informed" role must be informed about a decision because they are affected. [IV.C.4]

39. a; The themes into which ideas are grouped in an affinity diagram are suggested by the ideas themselves. [IV.D.1]

40. c; If an issue is ill-defined or difficult to grasp, the best tool to use is an affinity diagram, in which brainstormed ideas are arranged into natural groups through a group consensus. [IV.D.1]

41. c; An affinity diagram can be used to organize the ideas developed during a brainstorming session into larger categories. [IV.D.1]

42. d; An organization chart is an example of a tree diagram. [IV.D.2]

43. d; Quality function deployment uses the house of quality, which is an example of a matrix diagram. [IV.D.3]

44. a; The matrix diagram compares two or more sets of data to discover existing relationships. [IV.D.3]

45. c; This matrix compares the pairwise relationships between four variables: division, part, customer, and shipper. It is a Y-shaped matrix. [IV.D.3]

46. c; The matrix shows that Customer 2 buys a large volume of Part B. We can also see that Customer 1 does not order Part B, and that Division 2 produces only Part B and ships it using only Shipper 1. [IV.D.3]

47. c; To compare the pairwise relationships between three variables, use a T-shaped matrix. [IV.D.3].

48. c; We calculate the sum of the weighted ranks as shown below. For example, the total rank for each product is calculated as:

$$Rank\,A = (4\times0.50)+(3\times0.25)+(1\times0.25) = 3.00$$
$$Rank\,B = (3\times0.50)+(1\times0.25)+(2\times0.25) = 2.25$$
$$Rank\,C = (1\times0.50)+(2\times0.25)+(4\times0.25) = 2.00$$
$$Rank\,D = (2\times0.50)+(6\times0.25)+(3\times0.25) = 2.75.$$

	Ease of Use	Low Cost	Time to Install	
Product	**50%**	**25%**	**25%**	**Total Rank**
A	4	3	1	3.00
B	3	1	2	2.25
C	1	2	4	2.00
D	2	4	3	2.75

Since the ranking system has 1 = the best, we look for the product with the lowest total rank. Product C should be considered first. [IV.D.4]

49. c; The activity network with durations is shown below. There are three paths: ACF = 19, BDF = 18, and EF = 21. EF is the path with the longest duration and is therefore the critical path. [IV.D.5]

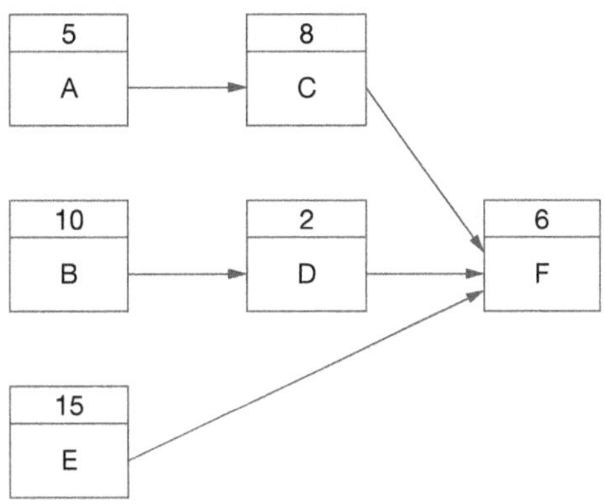

50. a; The full analysis of the activity diagram is shown below. For each activity, the duration is listed at the top. Under each activity, the early start and early finish appear. Under those figures, the late start and late finish are shown. The early start for Activity C is then equal to 5, shown circled below. [IV.D.5]

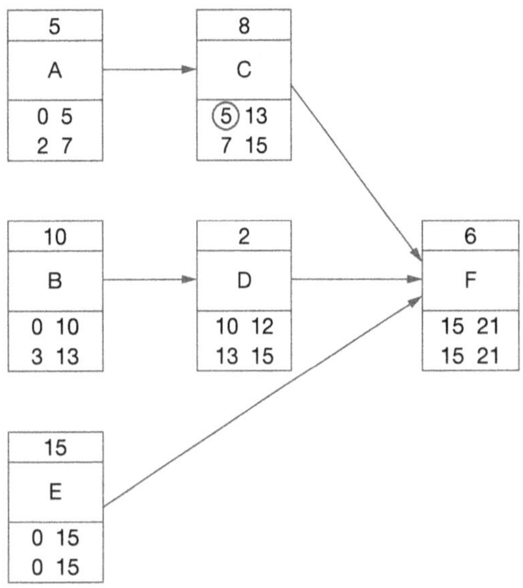

51. d; Because we have identified the critical path to be EF, if activity E is delayed by five days, the entire project will be delayed by five days even if all other activities are completed on schedule. [IV.D.5]

52. b; The late finish for an activity in an activity network diagram shows the latest that the activity can finish without delaying the project. For activities on the critical path, early finish and late finish values are equal. [IV.D.5]

53. b; By definition, all activities in the critical path have slack values equal to zero. If any of these activities takes longer to complete, the entire project will be delayed. [IV.D.5]

54. d; The process decision program chart (PDPC) is used to list all the ways an activity can go wrong and then propose countermeasures in the event of a problem. [IV.D.6]

55. b; A process decision program chart begins as a tree diagram. The team then brainstorms what could go wrong for each third-level task and then develops counter measures. [IV.D.6]

56. a; The nodes with the most outgoing arrows are considered drivers or causes of the problem. The nodes with the most incoming arrows are considered outcomes or effects. [IV.D.7]

Section V

Measure

V. Measure (25 Questions)

A. ***Process characteristics***

1. *Process flow metrics*. Identify and use process flow metrics (e.g., work in progress (WIP), work in queue (WIQ), touch time, takt time, cycle time, and throughput) to determine constraints. Describe the impact that "hidden factories" can have on process flow metrics. (Analyze)

2. *Process analysis tools*. Select, use, and evaluate various tools, e.g., value stream maps, process maps, work instructions, flowcharts, spaghetti diagrams, and gemba walk. (Evaluate)

B. ***Data collection***

1. *Types of data*. Define, classify, and distinguish between qualitative and quantitative data, and continuous and discrete data. (Evaluate)

2. *Measurement scales*. Define and use nominal, ordinal, interval, and ratio measurement scales. (Apply)

3. *Sampling*. Define and describe sampling concepts, including representative selection, homogeneity, bias, accuracy, and precision. Determine the appropriate sampling method (e.g., random, stratified, systematic, subgroup, and block) to obtain valid representation in various situations. (Evaluate)

4. *Data collection plans and methods*. Develop and implement data collection plans that include data integrity, accuracy, and processing tools, e.g., check sheets and data normalization. Avoid data collection pitfalls by defining the metrics to be used or collected, ensuring that collectors are trained in the tools and understand how the data will be used, and checking for seasonality effects. (Analyze)

C. *Measurement systems*

1. *Measurement system analysis (MSA).* Use gauge repeatability and reproducibility (R&R) studies and other MSA tools (e.g., bias, correlation, linearity, precision to tolerance, and percent agreement) to analyze variable measurement system capability. Use audit MSA for attribute measurement systems. (Evaluate)

2. *Measurement systems across the organization.* Identify how measurement systems can be applied to all functional areas of the organization (e.g., marketing, sales, engineering, research and development (R&D), supply chain management, operations and customer experience). (Understand)

3. *Metrology.* Define and describe elements of metrology, including calibration systems, traceability to reference standards, and the control and integrity of measurement devices and standards. (Understand)

D. *Basic statistics*

1. *Basic statistical terms.* Define and distinguish between population parameters and sample statistics, e.g., proportion, mean, and standard deviation. (Apply)

2. *Central limit theorem.* Explain the central limit theorem and its significance in the application of inferential statistics for confidence intervals, hypothesis tests, and control charts. (Understand)

3. *Descriptive statistics.* Calculate and interpret measures of dispersion and central tendency. (Evaluate)

4. *Graphical methods.* Construct and interpret diagrams and charts, e.g., box-and-whisker plots, scatter diagrams, histograms, normal probability plots, frequency distributions, and cumulative frequency distributions. (Evaluate)

5. *Valid statistical conclusions.* Distinguish between descriptive and inferential statistical studies. Evaluate how the results of statistical studies are used to draw valid conclusions. (Evaluate)

E. *Probability*

1. *Basic concepts.* Describe and apply probability concepts, e.g., independence, mutually exclusive events, addition and multiplication rules, conditional probability, complementary probability, and joint occurrence of events. (Understand)

2. *Common distributions.* Describe, interpret, and use normal, Poisson, binomial, chi-square, Student's t, and F distributions. (Evaluate)

3. *Additional distributions.* Identify hypergeometric, bivariate, exponential, lognormal, and Weibull distributions. (Understand)

F. *Process capability*

1. *Process capability indices.* Define, select, and calculate C_p and C_{pk}. (Evaluate)
2. *Process performance indices.* Define, select, and calculate P_p, P_{pk}, C_{pm}, and process sigma. (Evaluate)
3. *General process capability studies.* Describe and apply elements of designing and conducting process capability studies relative to characteristics, specifications, sampling plans, stability, and normality. (Evaluate)
4. *Process capability for attributes data.* Calculate the process capability and process sigma level for attributes data. (Apply)
5. *Process capability for non-normal data.* Identify non-normal data and determine when it is appropriate to use Box-Cox or other transformation techniques. (Apply)
6. *Process performance versus specification.* Distinguish between natural process limits and specification limits. Calculate process performance metrics, e.g., percent defective, parts per million (PPM), defects per million opportunities (DPMO), defects per unit (DPU), first pass yield, and rolled throughput yield (RTY). (Evaluate)
7. *Short-term and long-term capability.* Describe and use appropriate assumptions and conventions when only short-term data or only long-term data are available. Interpret the relationship between short-term and long-term capability. (Evaluate)

QUESTIONS

1. Which of the following items are components of work in progress (WIP)?
   a. Work in queue (WIQ) and work being processed
   b. Finished goods and inventory
   c. Work being processed and currently shipped
   d. Work in queue (WIQ) and parts in transit
2. The time that a unit of product is being worked on at any step in the process is defined as:
   a. cycle time.
   b. set up time.
   c. touch time.
   d. takt time.
3. The amount of output that passes through the process in a specified period is defined as:
   a. touch time.
   b. cycle time.
   c. takt time.
   d. throughput.
4. A customer requires 320 units per day. The plant operates continuously for one eight-hour shift. What is the required takt time in minutes per unit?
   a. 0.67
   b. 1.5
   c. 40
   d. 320
5. When a process step changes the form, fit, or function of a product, this is referred to as:
   a. work in queue (WIQ).
   b. work in progress (WIP).
   c. eight wastes.
   d. value-added activity.

6. The time from when a hospital patient is discharged to when that hospital room is ready for the next patient is known as the:

    a. value-added time.

    b. cycle time.

    c. takt time.

    d. set-up time.

7. The collection of activities in a process that generates waste is called the:

    a. hidden factory.

    b. process map.

    c. swim lane chart.

    d. process performance.

8. A chart that graphically represents the steps needed to approve a loan application would be referred to as a:

    a. scatter plot.

    b. why-why diagram.

    c. process flow chart.

    d. interrelationship digraph.

9. A value stream map (VSM) is a detailed process map that shows what types of information?

    a. Stakeholders, current state, and action plans

    b. WIP, changeover times, and inventory

    c. Defect locations and error types

    d. Material and worker flow and distances traveled

10. What does the following symbol on a value stream map represent?

    a. Safety or buffer stock

    b. Production kanban

    c. Inventory

    d. Supermarket

11. What does the following symbol on a value stream map represent?

    a. Safety or buffer stock
    b. Defect detection
    c. Kaizen burst
    d. Supermarket

12. Work instructions should ideally:
    a. be printed on A3-sized paper.
    b. provide a broad overview of process steps.
    c. should encourage improvisation.
    d. include labeled graphics.

13. A team would like to track the physical path that parts travel from arrival at the dock through the manufacturing process. Which graphical tool would be the best choice?
    a. Value stream map
    b. Spaghetti diagram
    c. Gemba walk
    d. Process map

14. Before creating a process map, a team should:
    a. brainstorm ideas to improve the process.
    b. go on a gemba walk.
    c. hold stakeholders accountable.
    d. create a future state map.

15. A flow chart that can show handoffs and changes of control across departments is known as a:
    a. swim lane chart.
    b. value stream map.
    c. spaghetti diagram.
    d. SIPOC diagram.

16. Which of the following characteristics is considered qualitative?
    a. Weight
    b. Length
    c. Volume
    d. Defect type
17. Which of the following measurements is considered discrete data?
    a. The weight of a raw material
    b. The number of defects per unit
    c. The output voltage of a transformer
    d. The cycle time for a filling operation
18. Data collected on which measurement scale is classified into categories with no order implied?
    a. Ordinal
    b. Interval
    c. Ratio
    d. Nominal
19. In this type of measurement scale, data can be arranged in a series in which order is important, but precise differences between values are not defined.
    a. Interval
    b. Ratio
    c. Ordinal
    d. Nominal
20. This measurement scale has meaningful differences, and an absolute zero exists.
    a. Interval
    b. Ratio
    c. Ordinal
    d. Nominal

21. What is an acceptable measure of dispersion to use for data that is on the ordinal measurement scale?

    a. Standard deviation

    b. Interquartile range

    c. Mode

    d. None is acceptable

22. A company creates a customer satisfaction survey using a five-point Likert scale. Which measurement scale is associated with this type of data?

    a. Nominal

    b. Ordinal

    c. Interval

    d. Ratio

23. An injection molding process has a four-cavity mold running in one press. What sampling method is most appropriate to monitor this process?

    a. Double sampling

    b. Simple random sampling

    c. Stratified sampling

    d. Multiple sampling

24. An internet poll reveals that respondents prefer Product A over Product B. This poll result is likely not representative of the population due to:

    a. recall bias.

    b. self-selection bias.

    c. survivorship bias.

    d. exclusion bias.

25. A Black Belt is preparing a simple random sample. List the steps, from first to last, that they must perform to take the sample.

    1. Select the members of the population associated with the n random numbers.

    2. Determine the sampling frame.

    3. Generate a random sample of n different numbers between 1 and N.

    4. Assign each member of the population a number from 1 to N.

a. 3, 2, 4, 1

b. 2, 4, 1, 3

c. 4, 3, 2, 1

d. 2, 4, 3, 1

26. A quality inspector chooses every 10th assembly coming off the final production line for testing. What type of sampling scheme are they using?

a. Simple random sampling

b. Systematic sampling

c. Convenience sampling

d. Stratified sampling

27. A company sends a survey to its 2400 customers. What is the minimum number of survey responses required to achieve a margin of error equal to ± 3.5%?

a. 28

b. 29

c. 609

d. 610

28. What is the purpose of data normalization?

a. To transform the data into a normal distribution

b. To put different data sets on a common measurement scale

c. To ensure that units are assigned to data values

d. To detect sampling bias

29. A useful tool to keep a tally of events is known as a:

a. Gantt chart.

b. fishbone diagram.

c. check sheet.

d. Y-matrix.

30. A data collection plan for seasonal data should:

a. capture data within several seasonal cycles.

b. use as short a collection period as possible to avoid data contamination.

c. ignore seasonality.

d. collect data at regular intervals.

31. In measurement systems, the term *accuracy* encompasses which two components?

a. Bias and precision

b. Bias and linearity

c. Stability and linearity

d. Repeatability and precision

32. The stability of a measurement system refers to its:

a. accuracy.

b. reproducibility.

c. accuracy and precision over time.

d. repeatability over time.

33. In measurement system analysis, which of the following measures is a component of precision?

a. Stability

b. Linearity

c. Bias

d. Repeatability

34. A scale can weigh an assembly to the nearest pound. The weight specifications for the assembly are 5 kg ± 0.6 kg. What can be said about the instrument?

a. The scale is not a precise instrument.

b. The scale cannot discriminate between conforming and nonconforming assemblies.

c. The scale is biased toward accepting defective assemblies.

d. The scale is not a reproducible instrument.

35. A precision-to-tolerance ratio is calculated to be 0.50. What can be said about the measurement system?

a. The measurement system is stable.

b. The variability of the measurement system is too large.

c. The variability of the products is too large.

d. The specification range is too small.

36. In a measurement system analysis, percent contribution is equal to:

a. the square root of the precision.

b. the square of the discrimination.

c. the precision-to-tolerance ratio.

d. the square of the total percent gage R&R.

37. If the bias of a measuring system increases across the measurement range, the measuring device is said to have poor:

a. stability.

b. linearity.

c. bias.

d. reproducibility.

Use the figure below to answer questions 38 and 39.

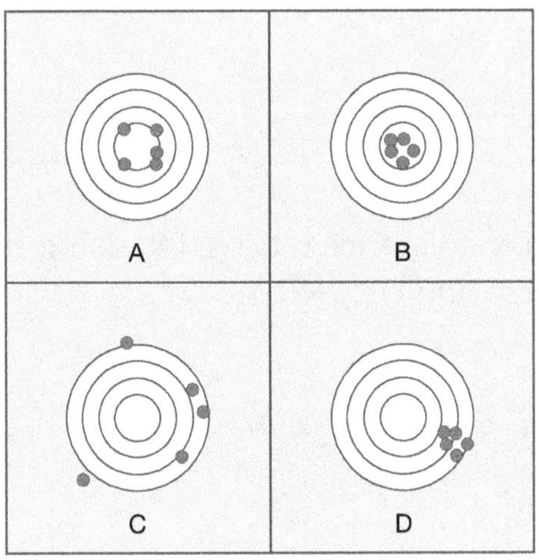

38. Which quadrant of the figure shows high precision but low accuracy?

a. Quadrant A

b. Quadrant B

c. Quadrant C

d. Quadrant D

39. Which quadrant in the figure shows high accuracy and low precision?
    a. Quadrant A
    b. Quadrant B
    c. Quadrant C
    d. Quadrant D

40. What is the measurement system concept depicted in the figure below?

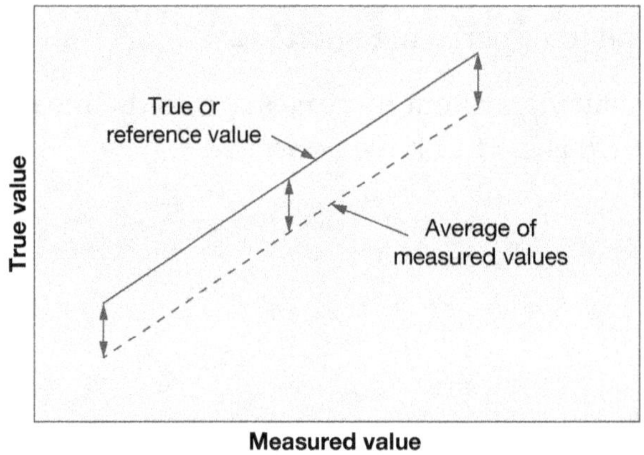

    a. Precision
    b. Stability
    c. Linear bias
    d. Repeatability

41. Calculate the precision-to-tolerance ratio (PTR) using the following information: Specifications 10 ± 0.5 cm and $\sigma_{total} = 0.045$ cm.
    a. 13.5%
    b. 27%
    c. 37%
    d. 75%

42. A measurement system has been evaluated with the following results: repeatability = 0.81 and reproducibility = 5.15. What is the gage R&R?
    a. 2.44
    b. 5.21
    c. 5.96
    d. 27.14

43. The gage R&R for a measurement system is equal to 3.45. The equipment variation is found to be equal to 1.22. What is the appraiser variation?

    a. 1.40

    b. 2.23

    c. 3.23

    d. 3.27

44. The standard deviation due to repeatability is also known as:

    a. gage R&R.

    b. equipment variation.

    c. appraiser variation.

    d. precision to tolerance.

45. The number of distinct categories for a measurement system is equal to 7. If the gage R&R is equal to 0.40, what is the value for σ_{parts}?

    a. 0.79

    b. 1.98

    c. 3.96

    d. 247.5

46. A single operator measures 30 parts two times each. The average range of the readings is equal to 2.45 mm. What is the standard deviation due to repeatability?

    a. 1.45

    b. 2.09

    c. 2.17

    d. 4.71

47. Two operators measure three parts two times each. The average measurement is 45.67 cm for Operator 1 and 47.54 cm for Operator 2. The equipment variation is equal to 1.25 cm. What is the appraiser variation?

    a. 1.03

    b. 1.22

    c. 1.49

    d. 1.54

48. What components comprise the total variation of a measurement system?

    a. Gage R&R and appraiser variation

    b. Part-to-part variation and gage R&R

    c. Equipment variation and gage R&R

    d. Part-to-part variation and repeatability

49. ABC Manufacturing performed a gage R&R study on a pair of calipers used to measure the width of a part. The customer has since changed the specifications on the drawing for the width. What part of the gage R&R is likely affected when a new gage R&R study is performed?

    a. Percent study variation

    b. Repeatability

    c. Percent tolerance

    d. Reproducibility

50. What method can be used to assess a measurement system that uses pass/fail classification?

    a. AIAG methodology

    b. ANOVA analysis

    c. Control chart method

    d. Percent agreement

51. A gage R&R study indicates that the number of distinct categories is less than five. Which of the following steps should be taken next?

    a. Repeat the gage R&R study using the same samples and operators.

    b. Reduce the number of samples.

    c. Reduce the number of operators.

    d. Review the gage for improvement.

52. Attribute agreement analysis is conducted by rating appraisers on what type of scale?

    a. Nominal

    b. Ordinal

    c. Interval

    d. Ratio

53. An attribute agreement analysis can include output on:
    a. each appraiser versus standard and part variation.
    b. all appraisers versus the standard and between appraiser agreement.
    c. within appraiser agreement and part variation.
    d. all appraisers versus standard and gage R&R.

54. A torque wrench is verified to be within specifications with a torque analyzer before every use at ABC Manufacturing. ABC wants to move to verifying the torque wrench once a week. What measurement systems analysis evaluation technique will ensure that the torque wrench will exhibit only common cause variation over the week?
    a. Bias
    b. Linearity
    c. Stability
    d. Reproducibility

55. The range of weights for five parts is calculated to be 0.75 grams. What is the part variation?
    a. 0.091
    b. 0.112
    c. 0.302
    d. 0.335

56. What is the total variation of the measurement system that has part variation equal to 1.60, EV = 0.75, and gage R&R equal to 0.89?
    a. 1.83
    b. 2.49
    c. 2.65
    d. 3.35

57. Audit MSA is concerned with:
    a. assessing the reliability of audit results.
    b. investigating where the data comes from, who collects it, and how the data was generated.
    c. assessing the validity of reproducibility findings.
    d. investigating the database used to store measurement readings.

58. The percent tolerance for a measurement system is calculated to be equal to 30%. The measurement system can be described as:

    a. inadequate.

    b. marginal.

    c. acceptable.

    d. failed.

59. A gage R&R study is conducted with three inspection technicians. In the analysis, how many degrees of freedom (df) are associated with the technicians?

    a. 1

    b. 2

    c. 3

    d. 4

60. Measurement systems:

    a. are only used in manufacturing applications.

    b. cannot be applied to human resources.

    c. can be used to assess supplier performance.

    d. are separate from the marketing function.

61. An example of a measurement system used in human resources is:

    a. customer loyalty surveys.

    b. performance evaluations.

    c. equipment calibration.

    d. gage traceability.

62. A gage traceability document:

    a. lists common causes of measurement error.

    b. sets the calibration schedule.

    c. records variation in the measurement system.

    d. links a given measurement system to standards maintained by NIST.

63. The purpose of calibration is to:

    a. ensure that the measurement system is performing correctly.

    b. vet the measurement standard against the equipment.

c. calculate the gage R&R of an instrument.

d. adjust measurement devices to reduce defect detection.

64. An example of measurement error due to equipment is:

a. lack of precision.

b. lack of physical ability.

c. parallax effect.

d. vibration.

65. A one-time custom order of 75 commemorative coins was minted. The weight of each coin was recorded, and a Black Belt calculated the mean and standard deviation from the data. The calculated values are considered to be:

a. statistics.

b. parameters.

c. estimates.

d. ordinal measures.

66. What is the denominator in the population variance formula?

a. N

b. $n - 1$

c. $N - 1$

d. $\sqrt{N}$

67. A survey was conducted to gather employees' opinions on work locations. The preferences for each option are shown below:

Option	Preference
In-office full-time	12%
Hybrid	45%
Work from home full-time	33%

The percentages in the table above are considered to be:

a. proportions.

b. means.

c. intervals.

d. ranges.

68. Which statistic is a measure of central tendency?

   a. Sample variance
   b. Sample range
   c. Sample mode
   d. Sample correlation

69. Twenty samples are taken from a symmetric distribution with $\mu = 34.5$ and $\sigma^2 = 4.5$. What is the standard deviation of the sampling distribution of the average?

   a. 0.225
   b. 0.474
   c. 1.006
   d. 4.528

70. What theorem helps us assume that the plotted points on an X-bar chart are normally distributed?

   a. Bayes' theorem
   b. Central limit theorem
   c. Gauss-Markov theorem
   d. Law of large numbers

71. One hundred independent samples are taken from an exponential distribution with a mean equal to 2.0 and a variance equal to 4.0. The samples are summed. What is the approximate distribution of the sum of the values?

   a. Normal (2.0, 0.04)
   b. Normal (2.0, 0.40)
   c. Exponential (200)
   d. Normal (200, 400)

72. Forty samples are taken from an exponential distribution with parameter $\theta = 2$. Which distribution should be used to construct a 95% confidence interval on the mean?

   a. Normal
   b. Student's *t*
   c. Exponential
   d. Log normal

73. The measure of the heaviness of a distribution's tails is known as:

    a. kurtosis.

    b. skewness.

    c. variation.

    d. range.

74. The skewness of a distribution is equal to –2.3. What is the shape of the distribution?

    a. Bell-shaped

    b. Skewed to the right

    c. Skewed to the left

    d. Disjointed

Use the sample data {94, 91, 76, 43, 66, 77, 55, 27, 50, 60} to answer questions 75 and 76.

75. What is the value of the sample mean?

    a. 63.0

    b. 63.9

    c. 71.0

    d. 73.6

76. What is the value of the sample variance?

    a. 20.1

    b. 21.2

    c. 404.9

    d. 449.9

77. When a quick measure of dispersion is desired, which statistic can be used?

    a. Variance

    b. Standard deviation

    c. Range

    d. Median

78. Given the following sample data {155, 161, 152, 161, 153}, calculate the sample median.

    a. 152

    b. 153

    c. 155

    d. 161

79. Given the sample set of data {13.3, 14.5, –4.0, 23.9, 24.2}, calculate the standard deviation.

    a. 10.26

    b. 11.47

    c. 14.38

    d. Cannot be calculated.

80. Given the sample set of data {17.0, 14.5, 2.5, 4.5, 12.3}, calculate the standard error of the mean.

    a. 1.27

    b. 2.84

    c. 5.67

    d. 6.34

81. Calculate the mean using the frequency table shown below.

Bounds	Frequency
10–29	14
30–49	10
50–69	5
70–89	4

    a. 6.00

    b. 8.25

    c. 38.89

    d. 49.50

82. Calculate the interquartile range using the box plot shown below:

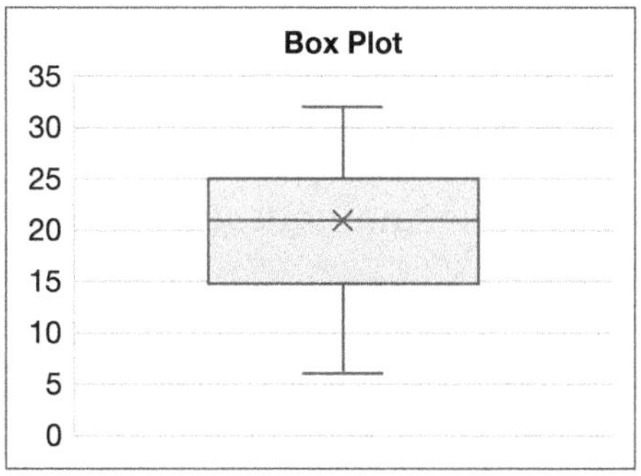

a. 10

b. 12

c. 21

d. 27

83. What insight does the box plot below show?

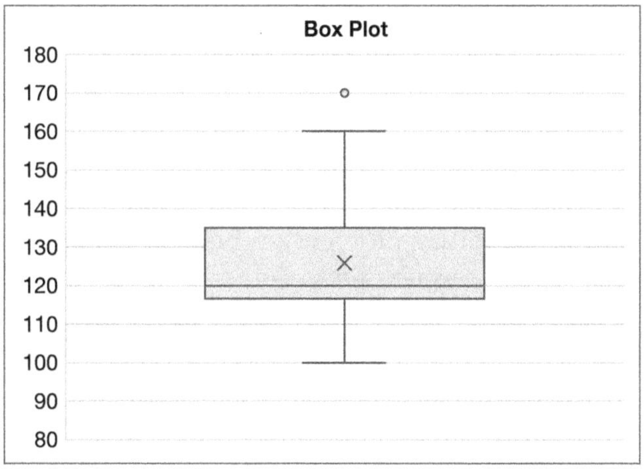

a. The data is symmetric.

b. The data is left skewed.

c. There is an outlier.

d. The median is greater than the mean.

84. Which of the following graphs can be used as a quick check for normality?

a. Histogram

b. Scatter plot

c. Run chart

d. Pareto chart

85. A scatter plot:

    a. shows the relationship between two quantitative variables.

    b. orders data from highest to lowest.

    c. lists causes such as materials, methods, and measurements.

    d. can only indicate linear relationships.

86. Given the frequency table below, what is the cumulative frequency associated with $X = 6$?

X	Frequency
4	10
5	12
6	14
7	4
8	2

    a. 14

    b. 22

    c. 36

    d. 42

87. Interpret the normal probability plot shown below.

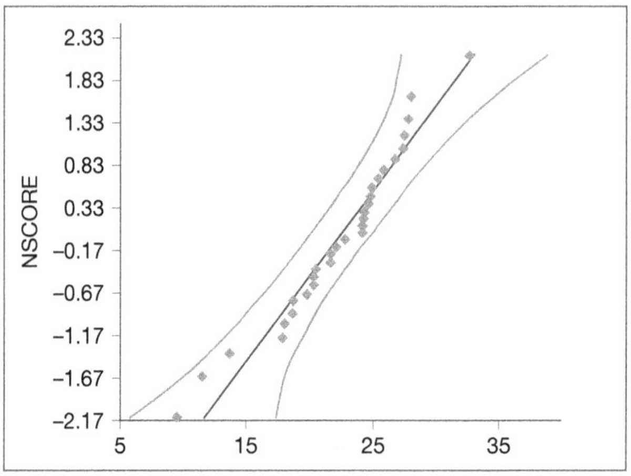

    a. The data are normally distributed.

    b. The data are right skewed.

    c. The standard deviation is greater than the mean.

    d. The data should be checked for errors.

88. Interpret the normal probability plot shown below.

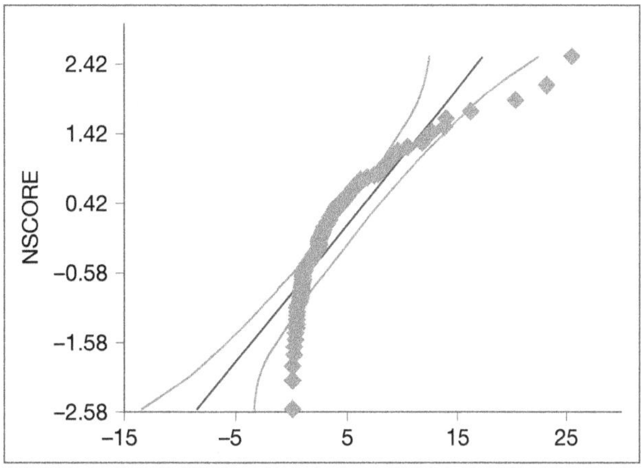

a. The data are normally distributed.

b. The data are right skewed.

c. The data are left skewed.

d. The data set has many negative values.

89. Metal plates are inspected for two types of defects: burrs and scratches. A Black Belt calculates the probability of finding a burr or a scratch on a metal panel. Which probability rule would they use?

a. Multiplication rule

b. Mutual exclusive rule

c. Complement rule

d. Addition rule

90. A project team consists of twelve members. How many ways can members be chosen for a three-member subcommittee?

a. 4

b. 12

c. 220

d. 1320

91. A sample of three numbers is selected from the numbers 1 to 45, inclusive. How many different combinations are possible?

a. 14,190

b. 28,380

c. 42,570

d. 85,140

92. When the occurrence of one event negates the possibility of another event, the events are referred to as:

a. independent.

b. mutually exclusive.

c. multiplicative.

d. empirical.

93. If events A and B are independent, which of the following statements is true about the conditional probability $\Pr(A|B)$?

a. $\Pr(A \mid B) = \Pr(A)$

b. $\Pr(A \mid B)$ cannot be determined.

c. $\Pr(A \mid B) = \Pr(A) + \Pr(B)$

d. $\Pr(A \mid B) = 0$

94. What is the main difference between a permutation and a combination?

a. For a given n and r, the number of combinations is generally larger than the number of permutations.

b. Order is important for combinations but not for permutations.

c. Order is important for permutations but not for combinations.

d. For a given n and r, the number of permutations is always greater than the number of combinations.

95. An event affects the probability of another event occurring. The events are referred to as:

a. independent.

b. multiplicative.

c. dependent.

d. conditional.

96. An inspector sampled four circuit boards from a steady stream of production that is known to be 12% nonconforming. What distribution would they use to find the probability of selecting two nonconforming units in the sample?

a. Binomial

b. Poisson

c. Normal

d. Hypergeometric

97. A customer service center receives an average of 2.5 calls per minute. What is the probability that the center receives fewer than 2 calls in the next minute?

a. 0.082

b. 0.257

c. 0.287

d. 0.544

98. Which of the following distributions is continuous?

a. Hypergeometric

b. Binomial

c. Poisson

d. Normal

99. An emergency department handles an average of 7.7 patients each hour on a Wednesday evening. How many arrivals should the check-in desk plan for if it wants to be staffed to handle the 95th percentile of arrivals?

a. 10

b. 11

c. 12

d. 14

100. A normally distributed process metric has a mean of 32.79 and standard deviation equal to 4.12. Four percent of production is scrapped because it falls outside the metric's upper specification limit. What is the value of the upper specification?

a. 40

b. 42

c. 45

d. 50

101. Ten square yards of carpet are inspected for cosmetic defects. The probability of finding two or fewer defects can be expressed using which probability statement shown below?

   a. $Pr(X = 2)$

   b. $Pr(X < 2)$

   c. $Pr(X \leq 2)$

   d. $1 - Pr(X = 2)$

102. A process output is normally distributed with $\mu = 53.7$ inches and $\sigma = 7.59$ inches. What percentage of product will be larger than 60 inches?

   a. 20.3%

   b. 25.2%

   c. 75.4%

   d. 80.0%

103. For normal distribution, a little more than two-thirds of the values are between:

   a. $\mu \pm 1.5\sigma$

   b. $\sigma \pm 1\mu$

   c. $\mu \pm 1\sigma$

   d. $\sigma \pm 1.5\mu$

104. A process is running at a 5% defective rate. An inspector samples 10 parts from the production line. What is the probability that they find fewer than three defectives?

   a. 0.0746

   b. 0.0105

   c. 0.9885

   d. 0.9990

105. Two manufacturing lines produce a bioplastic sheet with a tensile strength requirement of at least 9 MPa. The tensile strength for product from Line A is known to follow $N(12.5, 2.56)$. Tensile strength for Line B follows $N(11.5, 6.25)$. Which line has a lower yield?

   a. Line A

   b. Line B

c. Both have equal yields.

d. This cannot be determined from the information given.

106. Which distribution is used to test goodness of fit, association, and the standard deviation of the normal distribution?

a. Standard normal

b. *F*

c. Chi-square

d. Student's *t*

107. Which probability distribution is used for finite discrete data?

a. Hypergeometric

b. Poisson

c. Normal

d. Binomial

108. A bivariate normal distribution has how many parameters?

a. 2

b. 4

c. 5

d. 6

109. Which distribution is useful when modeling multiplicative data?

a. Exponential

b. Normal

c. Lognormal

d. Chi-square

110. An inspection procedure calls for sampling 5 items from a lot of 25. If no defective items are found, then the lot is accepted and shipped. What distribution would best model the probability of acceptance for a lot?

a. Hypergeometric

b. Binomial

c. Poisson

d. Normal

111. The Weibull distribution is useful when modeling what type of data?

   a. Reliability data

   b. Messy data

   c. Correlated data

   d. Defect counts

112. As parts are made, they are placed in trays that hold 30 parts. If a tray contains 6 defective parts and 2 parts are sampled from the tray, what distribution would be used to calculate the probability that both will be defective parts?

   a. Poisson

   b. Binomial

   c. Uniform

   d. Hypergeometric

113. Which statement is true concerning the memoryless property of the exponential distribution?

   a. The probability that $X < s$ is the same as the probability that $X > s$.

   b. The probability that a unit will function for at least 100 more hours given it has run for 1000 hours, or 2000 hours, is the same.

   c. The probability that a unit will function for 1000 hours, or 2000 hours, is the same.

   d. The probability that a unit will fail at time t is equal to 0.632.

114. What is the connection between the Weibull and exponential distributions?

   a. They are equivalent if the shape parameter equals 1.

   b. They are equivalent if the scale parameter is equal to 1.

   c. Both have an X range between $-\infty$ and ∞.

   d. The probability that X is less than the mean is 0.632.

115. A normally distributed, in-control, process output has a $C_p = 1.75$ and a $C_{pk} = 1.0$. What action should the team perform first to improve the process?

   a. Decrease the variability of the process.

   b. Move the process mean to the target.

   c. Increase the width of the specification limits.

   d. Perform a P_{pk} analysis.

116. Which of the following cases would be considered the most desirable in terms of the relationship of the specification to the process spread?

a. $6\sigma = USL - LSL$

b. $3\sigma = USL - LSL$

c. $6\sigma < USL - LSL$

d. $6\sigma > USL - LSL$

117. Of the following C_p values, which index is the most desirable?

a. 1.00

b. 1.50

c. 0.75

d. 0.30

118. Which of the following cases would result in a C_p index equal to 1?

a. $6\sigma < USL - LSL$

b. $6\sigma = USL - LSL$

c. $6\sigma > USL - LSL$

d. $3\sigma = USL - LSL$

119. The numerator of the C_p formula is also referred to as:

a. the voice of the customer.

b. the voice of the process.

c. the process precision.

d. the process variability.

120. What does a negative C_{pk} signify?

a. An error was made in the calculation.

b. The process mean is outside the specification limits.

c. The C_p value is less than 1.

d. The process variation is too large.

Use the following information to answer 121 and 122.

A process is monitored using an X-bar *S* chart and is in control with X-bar = 904.5, s-bar = 9.9, and n = 20. The specifications are 900+/−50.

121. What is the value of the C_p index?

a. 1.55

b. 1.66

c. 1.68

d. 1.75

122. What is the estimated natural tolerance?

a. 10.03

b. 59.40

c. 60.19

d. 100.0

Use the following information to answer questions 123 and 124.

A process metric is normally distributed, and an X-bar *R* chart shows that it is in-control with X-bar = 23.5, R-bar = 1.5, and n = 5. The specifications are 24+/−1.5.

123. What is the estimated C_{pk}?

a. 0.22

b. 0.52

c. 0.77

d. 1.03

124. What is the estimated C_{pm}?

a. 0.342

b. 0.572

c. 0.613

d. 3.78

125. Which statement is true about a process if C_{pk} = 0.85 and C_p = 1.10?

a. The process is capable and centered within the specification limits.

b. The process is centered within the specification limits but is not capable.

c. The process is not centered and not capable.

d. The process is narrow enough to be capable but is not centered.

126. The C_p, C_{pk}, and C_{pm} indices are used to characterize:

a. short-term capability.

b. long-term capability.

c. the 1.5 sigma shift.

d. the process sigma.

127. What information does the C_{pm} index incorporate that C_{pk} does not?

a. The location of the process mean relative to the specifications.

b. The variability of the process.

c. The distance between the process mean and the process target.

d. The distance between the process variability and the specifications.

Use the information below to answer questions 128 and 129.

A quality engineer analyzes five months of data taken on a process output. The average is equal to 45.67 and the standard deviation is 0.75. Specifications are 47.0+/−3.0.

128. What is the value of $\hat{P}_p$?

a. 0.44

b. 0.75

c. 1.33

d. 1.50

129. What is the value of $\hat{P}_{pk}$?

a. 0.37

b. 0.74

c. 1.35

d. 2.69

130. A process has a C_p = 1.25. What is its corresponding short-term process sigma?

a. 1.50

b. 2.25

c. 3.00

d. 3.75

131. A process metric has specifications equal to 0.20+/−0.01. If the estimated P_p equals 1.23, what is the standard deviation of the process metric?

a. 0.00542

b. 0.00271

c. 0.01626

d. 0.04100

132. A capability study was conducted on a normally distributed process metric using 28 subgroups. Calculate a 95% lower bound on C_{pk} given that the estimated $C_{pk} = 1.35$.

a. 0.97

b. 1.03

c. 1.07

d. 1.67

133. One step in conducting a capability study is to:

a. change the specifications to improve the capability.

b. check the normality of the data.

c. choose only the metrics that are performing best.

d. use as few subgroups as possible.

134. Given the following information from an in-control a *p*-chart, calculate the process capability: $\sum np = 55, \sum n = 2800$.

a. 0.012

b. 0.020

c. 0.028

d. 0.040

135. What statement is true about process capability calculated from an attributes control chart?

a. It is calculated using the same method as from a variables control chart.

b. It is equal to the width of specification limits divided by the width of the control limits.

c. It should be greater than 1.33.

d. It is equal to the control chart average.

136. Given the following information from an in-control np chart, calculate the process capability. The $\sum np = 120, n = 200$, and number of subgroups = 20.

a. 0.03

b. 0.06

c. 3.0

d. 6.0

137. Four stable processes are monitored using *u*-charts, with the following capabilities:

Process	Capability
A	2.31
B	0.67
C	1.67
D	1.33

Which process has the most desirable capability?

a. Process A

b. Process B

c. Process C

d. Process D

138. A stable *c*-chart has 25 points plotted with the sum of defects equal to 38. What is the capability of the process?

a. 0.66

b. 1.52

c. 1.67

d. 2.52

139. What Box-Cox transformation is used to transform data from a Poisson to a normal distribution?

a. $Y = X^{0.50}$

b. $Y = \ln(X)$

c. $Y = X^2$

d. $Y = 1/X$

140. To calculate the process capability for nonnormal data, a Black Belt can first use which statistical tool to convert the data to the normal distribution?

    a. Fisher information

    b. The central limit theorem

    c. Johnson transformation

    d. Gauss-Markov theorem

141. Tolerance is defined as:

    a. the customers' willingness to accept defects.

    b. the difference between the upper and lower specification limits.

    c. six times the specification width.

    d. six times the process standard deviation.

142. A total of 345 units were inspected and 15 were found to be defective. What is the percentage defective?

    a. 0.0435%

    b. 4.35%

    c. 15%

    d. 23%

143. A total of 54,012 units were produced, and 453 were found to be inoperable. What is the ppm defective?

    a. 119

    b. 453

    c. 839

    d. 8387

144. Units are each inspected for four types of defects. A total of 522 units were inspected with the following results:

Defect Type	Tally
A	12
B	29
C	10
D	4

Calculate the defects per unit.

a. 0.026

b. 0.105

c. 37.96

d. 55

145. There are five different defect types that can be found on any one unit: weight, length, thickness, diameter, and tensile strength. Six hundred units were inspected, and 34 defects were found. What is the DPMO?

a. 6.8

b. 5667

c. 11,333

d. 56,665

146. A process is running at 0.04 defects per unit. What is the first pass yield (FPY)?

a. 0.04

b. 0.673

c. 0.961

d. Cannot be calculated from the information given

147. Three thousand parts are produced. Thirty-five were sent out for rework and six were scrapped. What is the first pass yield?

a. 0.9863

b. 0.9883

c. 0.9923

d. 0.998

148. A process requires three steps operated in series. The FPY for each step are equal to 0.99, 0.95, and 0.97, respectively. What is the rolled throughput yield (RTY)?

a. 0.912

b. 0.95

c. 0.97

d. 0.992

149. A 3-sigma process output is normally distributed and centered at the specification target. If the mean shifts upward by 1.5 standard deviations, what is the resulting yield?

    a. 95.45%

    b. 93.32%

    c. 99.34%

    d. 99.73%

150. Short-term capability:

    a. is a snapshot of the process taken under ideal conditions.

    b. captures both within and between subgroup variation.

    c. does not need the process to be in-control.

    d. offers the most realistic view of the actual process performance.

151. Long-term capability:

    a. can be expressed using the C_{pm} index.

    b. requires the process to be in-control.

    c. captures both within and between subgroup variation.

    d. is often larger in magnitude than short-term measures.

SOLUTIONS

1. a; Work in progress (WIP) includes all the material that has been input into the process but has not reached the output or finished stage. This includes material being processed, waiting to be processed, or stored as inventory at each step. Therefore, work in queue (WIQ) and work being processed are components of WIP. [V.A.1]

2. c; The time a unit of product is being worked on at any step in the process is defined as touch time. In contrast, the cycle time is the time required to complete one unit from the beginning of the process to the end. [V.A.1]

3. d; The amount of output that passes through the process in a specified period of time is known as throughput. [V.A.1]

4. b; Takt time is the rate in time per unit at which the process must complete units to meet customer demand. It is calculated as the ratio of the total time available divided by the number of units required by the customer. Here, we have eight hours of time, which is equal to 480 minutes, and a customer demand equal to 320 units per day.

$$Takt\,time = \frac{Time\,available}{Units\,required} = \frac{480 \text{ min}}{320 \text{ units}} = 1.5 \text{ min / unit}$$

   [V.A.1]

5. d; Value-added activities change the form, fit, or function of a product. [V.A.1]

6. d; The time it takes from when a hospital patient is discharged to when the hospital room is ready for the next patient is called the set-up, or changeover, time. In manufacturing, this would be the time from the last unit of one job to the first good unit of the next job produced at full speed. [V.A.1]

7. a; The collection of activities in a process that generates waste is known as the hidden factory. [V.A.1]

8. c; A process flow chart uses symbols to show process steps and their sequence. It can be used to document the steps of any process. [V.A.2]

9. b; A value stream map (VSM) is a detailed process map that shows all the steps in sequence. Information gathered and displayed on a VSM can include cycle time, changeover time, number of people, WIP, inventory packaging sizes, scrap rates, value-added versus non-value-added activities, and number of changeovers. [V.A.2]

10. d; The symbol shown below is used to denote a supermarket in a value stream map. [V.A.2]

11. c; The symbol below is used to denote a kaizen burst (also called a kaizen idea) on a value stream map. [V.A.2]

12. d; Work instructions should ideally include labeled graphics, as well as a listing of all parts required, consumable raw materials, tooling and fixtures needed, personnel required, potential safety hazards, potential opportunities for defects, the exact sequence of steps, and a document and revision number. [V.A.2]

13. b; The team should use a spaghetti diagram to show how parts physically move through the process. Although a gemba walk may be needed to construct a spaghetti diagram, it is not a graphical tool. [V.A.2]

14. b; Before creating a process map, the team should go and walk the process (go on a gemba walk) to see the process as it really exists and talk to the operators who work on the process. [V.A.2]

15. a; A flow chart that can show handoffs and changes of control across departments is known as a swim lane chart. [V.A.2]

16. d; Defect type is qualitative since it is expressed as text and is on a nominal scale. [V.B.1]

17. b; The number of defects per unit is a whole number and is considered discrete, or count, data. Discrete data takes on either a finite set of values or a countably infinite set of values such as the set of counting numbers. [V.B.1]

18. d; Nominal data can be classified into categories, but there is no implied order. Nominal data can include qualitative data as well as numerical data that is used as a label (e.g., social security numbers or part numbers). No meaningful mathematical operations such as a sum or average can be performed on nominal data. [V.B.2]

19. c; Ordinal data can be thought of as ranking data. It has intrinsic order, but the differences between values are not necessarily the same. For example, a runner who comes in first may finish a race in 30 minutes, the second-place runner may finish in 31 minutes, but the third-place runner may finish in 45 minutes. [V.B.2]

20. b; Ratio data has an implied order, meaningful and equidistant differences between values, and an absolute zero. Therefore, statements on ratios between data values can be made. For example, if Part A weighs 3 kg and Part B weighs 6 kg, Part B can be said to weigh twice as much as Part A. [V.B.2]

21. b; For ordinal data, acceptable measures of dispersion include the range and interquartile range. [V.B.2]

22. b; Likert scales using numerical scales are by definition ordinal. The scores are ordered, but the difference between scores is not consistent. Analysis can include calculating sample mode, median, range, or interquartile range. [V.B.2]

23. c; For the four-cavity mold, it is best to sample from each cavity to ensure that each cavity is represented. This is called stratified sampling. [V.B.3]

24. b; Self-selection bias occurs when individuals decide whether to participate in the study. Those who do choose to participate are often not representative of the population as a whole. [V.B.3]

25. d; The steps to taking a simple random sample are as follows: determine the sampling frame; assign each member of the population a number from 1 to *N*; generate a random sample of *n* different numbers from 1 to *N*; and select the members of the population associated with the *n* random numbers. [V.B.3]

26. b; Choosing every *n*th item in a progression is known as systematic sampling. [V.B.3]

27. d; When the population size *N* is a known quantity, we can calculate the minimum sample size required to achieve a specified margin of error by using

$$\lceil n = \frac{N}{1+NE^2}.$$

[V.B.4]

where *N* is population size and *E* is the margin of error, expressed as a decimal. The function is a ceiling function, which means that we always round the result up to the next highest whole number.

$$\lceil n = \frac{N}{1+NE^2} = \frac{2400}{1+2400(0.035)^2} = 609.14$$

$$\lceil n = 610$$

28. b; Data normalization is used to transform data onto a common scale. The normalized data is unitless and has a mean of zero and standard deviation equal to one. Note that data normalization does not make a data set normally distributed. [V.B.4]

29. c; A check sheet is one of the 7 Quality Tools. It is used to tally the occurrence of events over time. A listing of categories or events is written on the left-hand side, with the space next to each category used to place a tally mark each time that category occurs. It is often used to track the frequency of defect types. [V.B.4]

30. a; For seasonal data, the data collection plan should be designed so that several seasonal cycles are represented. Collecting data at regular intervals may cause the plan to miss seasonal fluctuations in the data. [V.B.4]

31. b; In measurement systems analysis, the term accuracy encompasses both bias and linearity. [V.C.1]

32. c; The stability of a measurement system refers to its accuracy and precision over time. The stability can be monitored using control charts. [V.C.1]

33. d; Reproducibility and repeatability are both components of precision in a measurement system. [V.C.1]

34. b; If a measurement instrument cannot measure to the desired resolution, it is said that it cannot discriminate between objects. In this case, a good assembly weighing 5.56 kg and a nonconforming assembly weighing 5.90 kg would both read as 6 kg. [V.C.1]

35. b; A precision-to-tolerance ratio (PTR) is a measure of the capability of the measurement system. PTR values less than 0.10 are considered good, while values greater than 0.30 require improvement in the measurement system in terms of reducing its variability. [V.C.1]

36. d; The percent contribution of the measurement system is the square of the total percent gage R&R. Note that the total percent gage R&R is the precision-to-tolerance ratio expressed as a percentage. [V.C.1]

37. b; If the bias of a measuring system increases across the measurement range, it is said to have poor linearity. A system with good linearity will have a constant bias across the range of measurement. [V.C.1]

38. d; The quadrant in the lower right is displaying an example of high precision. It is consistent but not accurate since the points are not at the target. [V.C.1]

39. a; The quadrant in the upper left is displaying a system with high accuracy, since the points are near the target, but low precision because their location is not consistent. [IV.C.1]

40. c; The figure depicts the concept of linear bias, in which the measured values are consistently biased across the range of measurement. [V.C.1]

41. b; To calculate the precision-to-tolerance ratio (PTR), we use

$$PTR = \frac{6\hat{\sigma}_{total}}{USL - LSL} = \frac{6 \times 0.045}{10.5 - 9.5} = 0.27 = 27\%.$$

[V.C.1]

42. b; To calculate the gage R&R, we use

$$\begin{aligned}\sigma_{Gage\ R\&R} &= \sqrt{\sigma^2_{repeatability} + \sigma^2_{reproducibility}} \\ &= \sqrt{0.81^2 + 5.15^2} \\ &= 5.21.\end{aligned}$$

[V.C.1]

Section V Solutions

43. c; We can use the formula for gage R&R and a little algebra to find appraiser variation:

$$\sigma_{Gage\,R\&R} = \sqrt{EV^2 + AV^2}$$
$$3.45 = \sqrt{1.22^2 + AV^2}$$
$$AV^2 = 3.45^2 - 1.22^2$$
$$AV = \sqrt{10.41} = 3.23.$$

[V.C.1]

44. b; Repeatability is also known as equipment variation. It measures the precision under conditions where independent measurement results are obtained with the same method on identical items by the same appraiser using the same equipment within a short period of time. [V.C.1]

45. b; The part-to-part variability can be calculated using the formula for NDC and a little algebra. The symbol under NDC signifies a floor function, in which we round down to the nearest whole number for any result greater than 1.

$$\lfloor NDC = \sqrt{2}\frac{\sigma_{parts}}{\sigma_{gage\,R\&R}}$$
$$\sigma_{parts} = \frac{NDC \times \sigma_{gage\,R\&R}}{\sqrt{2}} = \frac{7 \times 0.40}{\sqrt{2}} = 1.98$$

[V.C.1]

46. c; The standard deviation of repeatability (or EV) is calculated using

$$EV = \bar{R} \times K_1$$

$$\text{where } K_1 = 0.8862 \text{ for 2 trials}$$

$$EV = 2.45 \times 0.8862 = 2.171.$$

[V.C.1]

47. b; The appraiser variation is the same as the standard deviation of reproducibility. It is calculated using

$$AV = \sqrt{(\overline{X}_{DIFF} \times K_2)^2 - \frac{EV^2}{nr}}$$

where:

$\overline{X}_{DIFF} = \max(\overline{X}_1, \overline{X}_2) - \min(\overline{X}_1, \overline{X}_2)$

$\overline{X}_i$ = the average of values measured by Appraiser i

$K_2 = 0.7071$ for two appraisers

EV = the equipment variation

n = the number of parts

r = the number of trials

$$AV = \sqrt{(1.87 \times 0.7071)^2 - \frac{1.25^2}{3 \times 2}} = 1.22.$$

[V.C.1]

48. b; The total variation in the measurement system comprises gage R&R and part-to-part variation. Gage R&R comprises equipment and appraiser variation. [V.C.1]

49. c; The calipers used for measurement have not changed, so it is not expected that the components of gage R&R will be affected by the change in specifications. However, the percent tolerance ratio will be affected. [V.C.1]

50. d; The percent agreement method can be used to assess a measurement system that uses pass/fail classification. [V.C.1]

51. d; According to the Automotive Industry Action Group (AIAG), if the number of distinct categories is less than five, the instrument should be reviewed for improvement. [V.C.1]

52. b; In attribute agreement analysis, appraisers are rated on an ordinal scale. [V.C.1]

53. b; The output for an attribute agreement analysis can include four components: each appraiser versus the standard, between appraisers, all appraisers versus the standard, and within appraiser. [V.C.1]

54. c; Stability addresses the variation in a measurement system over time. [V.C.1]

55. c; To calculate part variation, we use

$$PV = R_{part} \times K_3$$
$$\text{where } K_3 = 0.4030 \text{ for five units}$$
$$PV = 0.75 \times 0.4030 = 0.302.$$

[V.C.1]

56. a; Total variation of the measurement system is equal to

$$TV = \sqrt{GRR^2 + PV^2}$$
$$= \sqrt{0.89^2 + 1.60^2} = 1.83.$$

[V.C.1]

57. b; Audit MSA is concerned with investigating where the data comes from, who collects it, and how the data was generated. [V.C.1]

58. b; If the percent tolerance is less than 10%, the measurement system is considered acceptable. If it is between 10% and 30%, inclusive, the measurement system is considered marginal. Measurement systems with percent tolerance greater than 30% are considered inadequate. [V.C.1]

59. b; The degrees of freedom are equal to 1 less than the number of technicians, calculated as (3 – 1) = 2. [V.C.1]

60. c; Measurement systems can be used to assess supplier performance. [V.C.2]

61. b; Performance evaluations are an example of a measurement system used in human resources. [V.C.2]

62. d; A gage traceability document shows the relationship between a given measurement system and the standards maintained by the U.S. government at the National Institute of Standards and Technology (NIST). [V.C.3]

63. a; The purpose of calibration is to ensure that the measurement system is performing correctly. [V.C.3]

64. a; Causes of measurement error can be classified under equipment, methods, appraiser, environment, parts, and measurement. Lack of precision is a cause of equipment measurement error. [V.C.3]

65. b; The Black Belt has weighed all the items in the population. The calculated values are thus considered population parameters. [V.D.1]

66. a; The population variance formula is

$$\sigma^2 = \frac{\sum_{i=1}^{N} (x_i - \mu)^2}{N}.$$

[V.D.1]

67. a; The percentages from the survey show the proportion of respondents who preferred each option. [V.D.1]

68. c; The sample mode is a statistic that is a measure of central tendency. [V.D.1]

69. b; The central limit theorem states the standard deviation of the sampling distribution of the average will be

$$\sigma_{\bar{X}} = \frac{\sigma}{\sqrt{n}} = \frac{\sqrt{4.5}}{\sqrt{20}} = 0.4743.$$

[V.D.2]

70. b; The central limit theorem (CLT) states that the average of independent samples taken from a population with a known mean and variance will be approximately normal. The approximation improves as the sample size increases. Since the X-bar chart plots averages from a population, we are afforded some protection against non-normality by the CLT. [V.D.2]

71. d; The central limit theorem states that the distribution of a *sum* of n independent variables with mean μ and variance σ^2 will be approximately normally distributed with mean $n\mu$ and variance $n\sigma^2$. Therefore, the distribution of the sum of 100 samples will be $N(200, 400)$. [V.D.2]

72. a; The confidence interval on the mean can be constructed using the normal distribution due to the central limit theorem. The distribution of X is exponential with a mean $\beta = 2$ and variance $\beta^2 = 4$. From the central limit theorem, we know that the sample mean will approximately follow a normal distribution with mean equal to 2 and a variance equal to 4/40, or 0.10. The approximation improves as the sample size increases. [V.D.2]

73. a; The measure of the heaviness of a distribution's tails is its kurtosis. This measure is often mistakenly thought to represent the peakedness of a distribution. [V.D.3]

74. c; A negative value of skew implies that the distribution is skewed to the left. [V.D.3]

75. b; The sample mean is calculated using

$$\bar{x} = \frac{\sum_{i=1}^{n} x_i}{n} = \frac{639}{10} = 63.9.$$

[V.D.3]

76. d; The sample variance is calculated using

$$s^2 = \frac{\sum_{i=1}^{n}(x_i - \bar{x})^2}{n-1} = \frac{\sum_{i=1}^{n} x_i^2 - n\bar{x}^2}{n-1} = \frac{44{,}881 - 10(62.9^2)}{9} = 449.9.$$

[V.D.3]

77. c; The range is a measure of dispersion that can be obtained with minimal calculation. [V.D.3]

78. c; To find the median, we sort the data from low to high and choose the middle value, as shown. [V.D.3]

{152, 153, 155, 161, 161}

79. b; The sample standard deviation can be calculated using

$$s=\sqrt{\frac{\sum_{i=1}^{n}(x_i-\bar{x})^2}{n-1}}=\sqrt{\frac{\sum_{i=1}^{n}x_i^2-n\bar{x}^2}{n-1}}=\sqrt{\frac{1559.99-5(14.38)^2}{4}}=11.47.$$

[V.D.3]

80. b; The standard error of the mean is calculated using

$$s_e=\frac{s}{\sqrt{n}}$$

$$s=\sqrt{\frac{\sum_{i=1}^{n}(x_i-\bar{x})^2}{n-1}}=\sqrt{\frac{\sum_{i=1}^{n}x_i^2-n\bar{x}^2}{n-1}}=\sqrt{\frac{677.04-5(10.16)^2}{4}}=6.34$$

$$s_e=\frac{s}{\sqrt{n}}=\frac{6.34}{\sqrt{5}}=2.84.$$

[V.D.3]

81. c; Given the frequency table, we first calculate the midpoints M_i of each class. The midpoint is equal to

$$M=LB+(UB-LB)/2.$$

The midpoint of the first class is calculated as

$$M_1=10+(29-10)/2=19.5.$$

The other midpoints are calculated similarly.

Bounds	Midpoint	Frequency
10–29	19.5	14
30–49	39.5	10
50–69	59.5	5
70–89	79.5	4

We then calculate a weighted average using

$$\bar{X} = \frac{\sum_{i=1}^{n} M_i f_i}{\sum_{i=1}^{n} f_i} = \frac{(19.5\times 14)+(39.5\times 10)+(59.5\times 5)+(79.5\times 4)}{14+10+5+4} = \frac{1283.5}{33} = 38.89.$$

[V.D.3]

82. a; The interquartile range is the difference between the third and first quartiles. The middle 50% of the data is contained between these two quartiles. The upper boundary of the box denotes Q_3 and the lower boundary denotes Q_1. (25 – 15) = 10. [V.D.4]

83. c; The box plot shows that the data is right skewed and that the mean is greater than the median. The data value of 170 is an outlier. [V.D.4]

84. a; A histogram shows the center, spread, and shape of a distribution and can be used as a quick check for normality. [V.D.4]

85. a; A scatter plot shows the relationship between two quantitative variables. Based on the scatter plot, the analyst can assess the nature and direction of the relationship and the correlation between the variables. [V.D.4]

86. c; The cumulative frequency at X = 6 is equal to 36, as shown in the table below. [V.D.4]

X	Frequency	Cumulative Frequency
4	10	10
5	12	22
6	14	36
7	4	40
8	2	42

87. a; This normal probability plot has points that roughly follow the 45-degree line, with no points outside the 95% confidence bands. We can assume the data is normally distributed. [V.D.4]

88. b; The normal probability plot shows that the data are not normally distributed. There are many data points near the value of zero, and then fewer at higher values. This is a right-skewed distribution. In fact, the plot was created using an exponential distribution. [V.D.4]

89. d; Let Event A = Burr and Event B = Scratch. When finding the probability of Event A OR Event B, we use the addition rule. [V.E.1]

90. c; Since these are committee assignments, order does not matter. We calculate 12C3 = 220 using

$$12C3 = \frac{12!}{(12-3)!3!} = 220.$$

[V.E.1]

91. a; We calculate 45C3 = 14,190 using

$$45C3 = \frac{45!}{(45-3)!3!} = 14,190.$$

[V.E.1]

92. b; Mutually exclusive events cannot occur at the same time. A light switch is either on or off; it can't be both. [V.E.1]

93. a; If Events A and B are independent, knowing if either event has occurred gives us no information about the occurrence of the other event. Therefore $\Pr(A|B) = \Pr(A)$. [V.E.1]

94. c; For permutations, we take into account the order of the selected units. For combinations, we do not consider order. For a given n and r, the number of permutations will be larger than the number of combinations if $r > 1$. However, if $r = 1$, the number of permutations equals the number of combinations. [V.E.1]

95. c; If Events A and B are dependent, the occurrence of Event A affects the probability of occurrence of Event B, and vice versa. [V.E.1]

96. a; The binomial distribution is used when there are two outcomes (e.g., pass/fail), there are a fixed number of independent trials, and the probability of success is constant from trial to trial. [V.E.1]

97. c; This is a Poisson problem since the call counts are discrete, and we are given an arrival rate of $\mu = 2.5$ calls/minute. We want the probability that the center receives fewer than two calls in the next minute. This is $\Pr(X < 2) = \Pr(X \le 1)$. Here, we solve using

$$\Pr(X < 2) = \Pr(X \le 1) = \sum_{x=0}^{1} \frac{e^{-\mu}\mu^x}{x!} = e^{-2.5} + 2.5e^{-2.5} = 0.287.$$

[V.E.2]

98. d; The normal distribution is continuous. The other distributions listed are discrete probability distributions. [V.E.2]

99. c; This is a Poisson distribution problem since the arrivals are discrete and we are given an arrival rate equal to 7.7 patients per hour. We will need to use a

Section V Solutions

cumulative Poisson table like the one in Appendix C.2 in *The ASQ Certified Six Sigma Black Belt Handbook*. Using the table, we find the mean (denoted as λ in the *Handbook*) of 7.7 in the first column and then read across the row until we find a probability that is close to 0.95. We see that 0.9496 occurs in the column for $X = 12$. This implies that $\Pr(X \leq 12) = 0.9496$. If the check-in desk is staffed to handle 12 patients per hour, they will be able to meet the demand 95% of the time. [V.E.2]

100. a; The process output follows a normal distribution with $\mu = 32.79$ and $\sigma = 4.12$. We need to find the value ***a*** such that $\Pr(Z < a) = 0.96$. We will then convert the Z back to the original scale. Using a cumulative standard normal table like the one Appendix D in *The ASQ Certified Six Sigma Black Belt Handbook*, we first find a probability in the interior of the table that is closest to 0.96. Inspection shows that $\Pr(Z < 1.75) = 0.9599$. Now we convert $Z = 1.75$ back to the original scale using

$$Z = \frac{X - \mu}{\sigma}$$

$$1.75 = \frac{X - 32.79}{4.12}$$

$$X = (1.75 \times 4.12) + 32.79 = 40.$$

[V.E.2]

101. c; This is an application of a Poisson distribution. For discrete probability distributions, the probability of finding "two or fewer" includes 2, so the probability statement is $\Pr(X \leq 2)$. [V.E.2]

102. a; We are asked to find $\Pr(X > 60)$ where X follows a normal distribution with mean = 53.7 and standard deviation = 7.59 inches. Using the complement rule, we can write $\Pr(X > 60) = 1 - \Pr(X < 60)$. Next, we will use a Z transformation and the cumulative normal table to solve:

$$Z = \frac{X - \mu}{\sigma} = \frac{60 - 53.7}{7.59} = 0.830.$$

Using the cumulative normal table, we find $\Pr(Z < 0.83) = 0.7967$. Then $\Pr(X > 60) = 1 - \Pr(X < 60) = 1 - \Pr(Z < 0.83) = 1 - 0.7967 = 0.2033$. [V.E.2]

103. c; For a normal distribution, 68.3% of the distribution is within one standard deviation of the mean, or $\mu \pm 1\sigma$. [V.E.2]

104. c; This is a binomial problem because we are classifying samples as defective or nondefective, the probability of a defective is constant, and we have a set number of independent trials. The probability that there are fewer than 3 defectives in a sample of 10 is the same as Pr(X ≤ 2). We can solve using the formula for the binomial probability function, or we can use a cumulative binomial table like the one in Appendix B.2 in *The ASQ Certified Six Sigma Black Belt Handbook.* Using the formula to solve, we note that n = 10, p = 0.02, and X = 0, 1, and 2:

$$\begin{aligned}\Pr(X \le 2) &= \sum_{x=0}^{2}\binom{n}{x}p^x(1-p)^{n-x} = \sum_{x=0}^{2}\binom{10}{x}0.05^x(0.95)^{10-x}\\ &= \binom{10}{0}0.05^0(0.95)^{10} + \binom{10}{1}0.05(0.95)^9 + \binom{10}{2}0.05^2(0.95)^8\\ &= 0.9885.\end{aligned}$$

Alternatively, we can find n = 10 in the first column of the cumulative binomial table and read down to X = 2 and then across to p = 0.05 to find Pr(X ≤ 2) = 0.9885. [V.E.2]

105. b; Yield is defined as the percentage of product that meets specifications. Here, the specification states that sheets must have a tensile strength of at least 9 MPa.

For Line A:

$$\Pr(X \ge 9) = \Pr\left(Z \ge \frac{9-12.5}{\sqrt{2.56}}\right) = \Pr(Z \ge -2.19)$$

For Line B:

$$\Pr(X \ge 9) = \Pr\left(Z \ge \frac{9-11.5}{\sqrt{6.25}}\right) = \Pr(Z \ge -1.00)$$

Without looking up the probabilities in the standard normal table, we know that Pr(Z > –2.19) will be greater than Pr(Z > –1.00). Therefore, Line A has a higher percentage yield, and Line B has the lower yield. [V.E.2]

106. c; The chi-square distribution is used to test goodness of fit, association, and the standard deviation of the normal distribution. [V.E.2]

107. a; The hypergeometric distribution is used to model discrete data from a finite population. [V.E.3]

108. c; The bivariate normal distribution has five parameters: $\mu_1, \mu_2, \rho, \sigma_1, \sigma_2$. [V.E.3]

109. c; The lognormal is used to model multiplicative data. [V.E.3]

110. a; The hypergeometric would be used to find Pr(X = 0) in this case because we are sampling from a finite population and characterizing items as pass or fail. [V.E.3]

Section V Solutions

111. a; The Weibull distribution is useful when modeling reliability data. The bathtub curve can be modeled using three Weibull distributions with varying parameters. [V.E.3]

112. d; The hypergeometric distribution is used when sampling from a finite population in which sample units are characterized as pass or fail. In this case, the hypergeometric parameters are $N = 30$, $D = 6$, and $n = 2$. [V.E.3]

113. b; A consequence of the memoryless property of the exponential distribution is that the probability of a unit functioning at least another s hours does not depend on how long the unit has already been functioning. This is written $\Pr(X > (t + s) | X > t) = \Pr(X > s)$. [V.E.3]

114. a; The Weibull distribution simplifies into an exponential distribution with parameter β when the Weibull shape parameter (α) equals 1. [V.E.3]

115. b; The process potential, or C_p value, shows that the process is capable. The C_p value is comparable to a 5.25 process sigma. The C_{pk} value is not equal to the C_p value, indicating that the process is not centered at target. The first step the team should take is to move the mean to target, thus making $C_{pk} = C_p = 1.75$. After that, the team can work on reducing variation to further improve the process. [V.F.1]

116. c; The C_p value is a larger-the-better metric. For C_p to be as big as possible, the denominator (6σ) should be smaller than the numerator ($USL - LSL$). [V.F.1]

$$C_p = \frac{(USL - LSL)}{6\sigma}$$

117. b; The C_p value is a larger-the-better metric. The most desirable C_p index from those listed is 1.50, which is comparable to a 4.5 sigma level process. [V.F.1]

118. b; If the process spread (6σ) is equal to the specification width ($USL - LSL$), then the C_p value will equal 1. [V.F.1]

$$C_p = \frac{(USL - LSL)}{6\sigma}$$

119. a; The numerator of the C_p formula, ($USL - LSL$), is known as the voice of the customer. The denominator (6σ) is known as the voice of the process. [V.F.1]

120. b; If the C_{pk} value is negative, the process mean is located outside the specifications. If the mean is greater than the USL, or less than the LSL, then the C_{pk} formula will return a negative value. [V.F.1]

$$C_{pk} = \min\left[\frac{(USL - \mu)}{3\sigma}, \frac{(\mu - LSL)}{3\sigma}\right]$$

121. b; We are given X-bar = 904.5, s-bar = 39.9, and n = 20. The specifications are 900+/−50. To estimate sigma from the X-bar and S chart, we use

$$\hat{\sigma} = \frac{\overline{s}}{c_4}$$

where c_4 is a tabulated value based on the subgroup size n. This value can be obtained from a source such as Appendix K.5 in *The ASQ Certified Six Sigma Black Belt Handbook*. For n = 20, c_4 = 0.9869. We then estimate C_p using

$$\hat{C}_p = \frac{(USL - LSL)}{6\hat{\sigma}} = \frac{(950 - 850)}{6 \times \left(\frac{9.9}{0.9869}\right)} = 1.66.$$

[V.F.1]

122. c; We are given X-bar = 904.5, S-bar = 39.9, and n = 20. The specifications are 900+/−50. To estimate sigma from the X-bar and S chart, we use

$$\hat{\sigma} = \frac{\overline{s}}{c_4}$$

where c_4 is a tabulated value based on the subgroup size n. This value can be obtained from a source such as Appendix K.5 in *The ASQ Certified Six Sigma Black Belt Handbook*. For n = 20, c_4 = 0.9869. We then estimate NT using

$$\widehat{NT} = 6\hat{\sigma} = 6 \times \left(\frac{9.9}{0.9869}\right) = 60.19.$$

[V.F.1]

123. b; We are given X-double-bar = 23.5, R-bar = 1.5, and n = 5. The specifications are 24+/−1.5. To estimate sigma from the X-bar and R chart, we use

$$\hat{\sigma} = \frac{\overline{R}}{d_2}$$

where d_2 is a tabulated value based on the subgroup size n. This value can be obtained from a source such as Appendix K.5 in *The ASQ Certified Six Sigma Black Belt Handbook*. For n = 5, d_2 = 2.326. We then estimate C_{pk} using

$$\hat{C}_{pk} = \min\left[\frac{(USL - \overline{\overline{X}})}{3\hat{\sigma}}, \frac{(\overline{\overline{X}} - LSL)}{3\hat{\sigma}}\right] = \min\left[\frac{(USL - \overline{\overline{X}})}{3\frac{\overline{R}}{d_2}}, \frac{(\overline{\overline{X}} - LSL)}{3\frac{\overline{R}}{d_2}}\right]$$

$$= \min\left[\frac{(25.5 - 23.5)}{3 \times \left(\frac{1.5}{2.326}\right)}, \frac{(23.5 - 22.5)}{3 \times \left(\frac{1.5}{2.326}\right)}\right] = \min(1.03, 0.52) = 0.52.$$

[V.F.1]

Section V Solutions

124. c; We are given X-double-bar = 23.5, R-bar = 1.5, and n = 5. The specifications are 24+/−1.5. To estimate sigma from the X-bar and R chart, we use

$$\hat{\sigma} = \frac{\overline{R}}{d_2}$$

where d_2 is a tabulated value based on the subgroup size n. This value can be obtained from a source such as Appendix K.5 in *The ASQ Certified Six Sigma Black Belt Handbook*. For n = 5, d_2 = 2.326. We then estimate C_{pm} using

$$\hat{C}_{pm} = \frac{(USL - LSL)}{6\sqrt{\hat{\sigma}^2 + (\overline{\overline{X}} - T)^2}} = \frac{(USL - LSL)}{6\sqrt{\left(\frac{\overline{R}}{d_2}\right)^2 + (\overline{\overline{X}} - T)^2}}$$

$$= \frac{(25.5 - 22.5)}{6\sqrt{\left(\frac{1.5}{2.326}\right)^2 + (23.5 - 24.0)^2}} = \frac{3.0}{6\sqrt{0.665875}} = 0.613.$$

[V.F.2]

125. c; The C_p value is less than 1.33, so the process is not capable. In addition, the C_{pk} index does not equal the C_p value, indicating that the process is not centered at target. [V.F.1]

126. a; The C_p, C_{pk}, and C_{pm} indices are used to characterize short-term capability. Long-term capability is measured using P_p or P_{pk}. [V.F.1]

127. c; The C_{pm} formula includes the distance between the process mean and the process target in its denominator. This is known as the bias of the process. Statistically speaking, the denominator is six times the square root of the mean square error. [V.F.2]

$$C_{pm} = \frac{(USL - LSL)}{6\sqrt{\sigma^2 + (\mu - T)^2}}$$

128. c; Using X-bar = 45.67, s = 0.75, LSL = 44.0, and USL = 50, we calculate

$$\hat{P}_p = \frac{(USL - LSL)}{6s} = \frac{(50 - 44)}{6 \times 0.75} = \frac{1}{0.75} = 1.33.$$

[V.F.2]

Section V Solutions

129. b; Using X-bar = 45.67, s = 0.75, LSL = 44.0, and USL = 50, we calculate

$$\begin{aligned}\hat{P}_{pk} &= \min\left[\frac{(USL-\bar{X})}{3s}, \frac{(\bar{X}-LSL)}{3s}\right]\\ &= \min\left[\frac{(50.0-45.67)}{3\times 0.75}, \frac{(45.67-44.0)}{3\times 0.75}\right]\\ &= \min[1.92, 0.74]\\ &= 0.74.\end{aligned}$$

[V.F.2]

130. d; To find the short-term process sigma from a C_p value, multiply the C_p value by a factor of 3. Here, $1.25\times 3 = 3.75$. [V.F.2]

131. b; To find the process standard deviation, we can use the formula for the estimated P_p and do a bit of algebra:

$$\begin{aligned}\hat{P}_p &= \frac{(USL-LSL)}{6s}\\ 1.23 &= \frac{(0.21-0.19)}{6s}\\ s &= \frac{0.02}{6\times 1.23} = 0.00271.\end{aligned}$$

[V.F.2]

132. b; To calculate a 95% lower confidence bound on C_{pk}, we note that n = 28, $\hat{C}_{pk} = 1.35$, and $z_{0.05}$ = 1.645. Then

$$\begin{aligned}C_{pk} &\geq \hat{C}_{pk} - z_\alpha\sqrt{\frac{1}{9n}+\frac{\hat{C}_{pk}^2}{2n-2}}\\ &\geq 1.35-1.645\sqrt{\frac{1}{9\times 28}+\frac{1.35^2}{2\times 28-2}}\\ &\geq 1.35-1.645\times 0.1942\\ &\geq 1.03.\end{aligned}$$

[V.F.3]

133. b; When conducting a process capability study, make sure the process is stable, use 25–30 subgroups for your data set, and then check the normality of the data. If the data is not normal, a transformation can be used before calculating capability. [V.F.3]

134. b; The process capability from a *p*-chart is reported as the centerline, or p-bar:

$$\overline{p} = \frac{\sum np}{\sum n} = \frac{55}{2800} = 0.0196 \cong 0.02.$$

[V.F.4]

135. d; For attributes control charts, the process capability is reported as the average of the control chart, e.g., $\overline{p},\overline{np},\overline{u},\overline{c}$. [V.F.4]

136. d; The process capability from an *np*-chart is reported as the centerline of the chart. Let *k* equal the number of subgroups. Then

$$\overline{np} = \frac{\sum_{i=1}^{k}(np)_i}{k} = \frac{120}{20} = 6.0.$$

[V.F.4]

137. b; The process capability for a *u*-chart is expressed as the average number of defects per unit. Therefore, Process B has the most desirable capability. [V.F.4]

138. b; For a process monitored using a *c*-chart, the capability is equal to the average number of defects:

$$\overline{c} = \frac{\text{Sum of defects}}{\text{Number of points plotted}} = \frac{38}{25} = 1.52.$$

This gives us the average number of defects per sample of size *n*. [V.F.4]

139. a; The Box-Cox transformation from Poisson to normal uses a square root transformation, which is the same as $Y = X^{0.50}$. [V.F.5]

140. c; The Johnson transformation is a technique to convert nonnormal data to a normal distribution. The process capability can then be calculated using the transformed data. [V.F.5]

141. b; The tolerance is the difference between the upper and lower specification limits. [V.F.6]

142. b; To find percentage defective, we use

$$\frac{\textit{Total number of defective units}}{\textit{Total number of units}} \times 100\% = \frac{15}{345} \times 100\% = 4.35\%.$$

[V.F.6]

143. d; To calculate ppm defective, we use

$$ppm = \frac{\textit{Defectives}}{\textit{Total}} \times 10^6 = \frac{453}{54012} \times 10^6 = 8387.$$

[V.F.6]

144. b; To calculate defects per unit, we total up the defects and divide by the total number of units inspected. [V.F.6]

Defect Type	Tally
A	12
B	29
C	10
D	4
Total	**55**

$$dpu = \frac{Total\,defects}{Total\,units} = \frac{55}{522} = 0.105$$

145. c; To calculate defects per million opportunities (DPMO), we use

$$DPMO = \frac{Total\,defects}{Total\,opportunities} \times 10^6 = \frac{34}{600 \times 5} \times 10^6 = 11,333.$$

[V.F.6]

146. c; To find the first pass yield (FPY) from dpu, we can use the Poisson distribution to find Pr(X = 0). The mean of the Poisson distribution is dpu = 0.04. [V.F.6]

$$FPY = \Pr(X = 0) = e^{-\mu} = e^{-0.04} = 0.961$$

147. a; To find first pass yield (FPY), we can use

$$FPY = \frac{Total\,parts\,produced - rework - scrap}{Total\,parts\,produced} = \frac{3000 - 35 - 6}{3000} = 0.9863.$$

[V.F.6]

148. a; To find rolled throughput yield (RTY), we use

$$RTY = \prod_{i=1}^{n} FPY_i = 0.99 \times 0.95 \times 0.97 = 0.912.$$

[V.F.6]

149. b; For a 3-sigma process output that is normally distributed, the yield will equal 93.32% if the mean shifts upward by 1.5 standard deviations. This information is tabulated in several sources, such as *The ASQ Certified Six Sigma Black Belt Handbook*. [V.F.7]

150. a; Short-term capability is measured using C_p, C_{pk}, and C_{pm}. The indices provide a snapshot of the process under ideal conditions. [V.F.7]

151. b; Long-term capability measures such as P_p and P_{pk} capture both within and between subgroup variation. The measures are often calculated when the process is not in control. It is argued that these measures offer a more realistic view of the actual process performance. [V.F.7]

Section VI

Analyze

VI. Analyze (22 Questions)

A. ***Measuring and modeling relationships and variables***

1. *Correlation coefficient.* Calculate and interpret the correlation coefficient and its confidence interval and describe the difference between correlation and causation. (Evaluate)
2. *Linear regression.* Calculate and interpret regression analysis and apply and interpret hypothesis tests for regression parameters. Use the regression model for estimation and prediction, analyze the uncertainty in the estimate, and perform a residual analysis to validate the model. (Evaluate)
3. *Multivariate tools.* Understand sources of variation through multivariate (e.g., factor analysis, discriminant analysis, multiple analysis of variance (MANOVA)). (Understand)

B. ***Hypothesis testing***

1. *Terminology.* Define and interpret the significance level, power, type I, and type II errors of statistical tests. (Evaluate)
2. *Statistical versus practical significance.* Define, compare, and interpret statistical and practical significance. (Evaluate)
3. *Sample size.* Calculate sample size for common hypothesis tests: equality of means and equality of proportions. (Apply)
4. *Point and interval estimates.* Define and distinguish between confidence and prediction intervals. Define and interpret the efficiency and bias of estimators. Calculate tolerance and confidence intervals. (Evaluate)
5. *Tests for means, variances, and proportions.* Use and interpret the results of hypothesis tests for means, variances, and proportions. (Evaluate)
6. *Analysis of variance (ANOVA).* Select, calculate, and interpret the results of ANOVAs. (Evaluate)
7. *Goodness-of-fit (chi-square) tests.* Define, select, and interpret the results of these tests. (Evaluate)

C. *Risk analysis and management*

1. *Types of risk.* Identify, assess, and prioritize various types of risk such as enterprise, operational, supplier, security, product, and cybersecurity. (Analyze)

2. *Failure mode and effects analysis (FMEA).* Describe the purpose and elements of FMEA, including risk priority number (RPN), and evaluate FMEA results for processes, products, and services. Distinguish between design FMEA (DFMEA) and process FMEA (PFMEA) and interpret their results. (Evaluate)

D. *Additional analysis methods*

1. *Gap analysis.* Analyze scenarios to identify performance gaps and compare current and future states using predefined metrics. (Analyze)

2. *Root cause analysis.* Define and describe the purpose of root cause analysis, recognize the issues involved in identifying a root cause, and use various tools (e.g., 5 whys, Pareto charts, fault tree analysis, cause-and-effect diagrams, and A3) to resolve chronic problems. (Analyze)

3. *Waste analysis.* Identify and interpret the seven classic wastes (overproduction, inventory, defects, over-processing, waiting, motion, and transportation) and resource under-utilization. (Analyze)

QUESTIONS

Use the scatter plots below to answer questions 1 and 2.

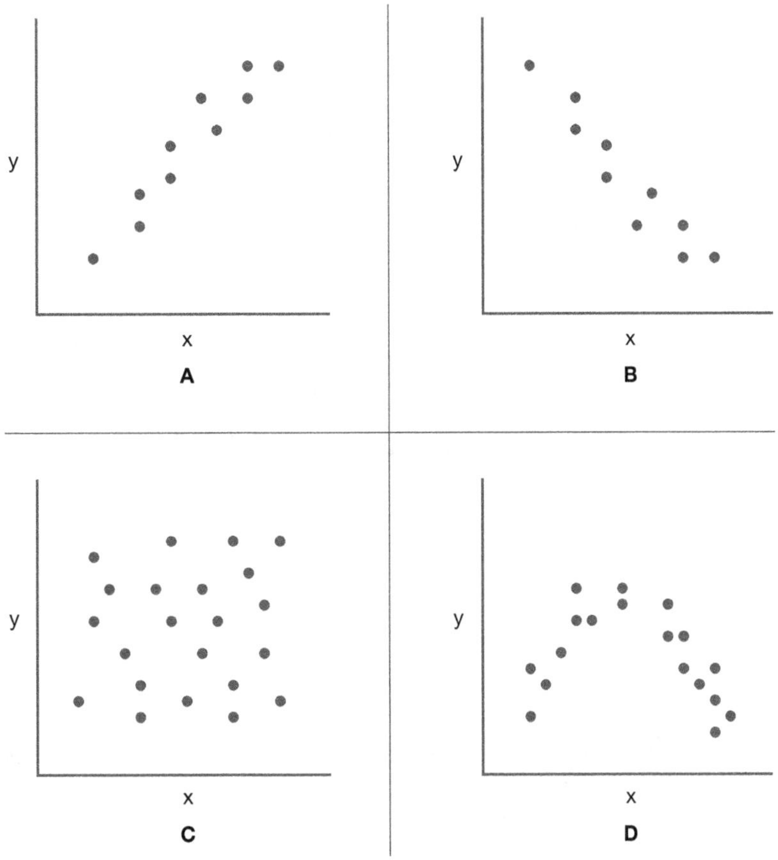

1. Which figure demonstrates negative correlation?
   a. Figure A
   b. Figure B
   c. Figure C
   d. Figure D
2. Which figure demonstrates a nonlinear relationship?
   a. Figure A
   b. Figure B
   c. Figure C
   d. Figure D

Use the table below to answer questions 3 and 4.

	Time	Strength			
	X	Y	XY	X^2	Y^2
	0.5	37.0	18.5	0.25	1369.00
	1.5	36.0	54.0	2.25	1296.00
	2.0	42.0	84.0	4.00	1764.00
	2.5	39.0	97.5	6.25	1521.00
	3.0	46.0	138.0	9.00	2116.00
	3.0	43.0	129.0	9.00	1849.00
	3.5	51.0	178.5	12.25	2601.00
	4.0	50.0	200.0	16.00	2500.00
SUM	**20.0**	**344.0**	**899.5**	**59.00**	**15016.00**

3. What is the value of the sample correlation coefficient?
   a. –0.96
   b. 0.77
   c. 0.88
   d. 0.96

4. What is the estimated slope of the regression line predicting strength from time?
   a. 1.91
   b. 2.10
   c. 4.39
   d. 14.33

5. A Black Belt calculates a sample correlation to be –0.37 using $n = 24$ pairs of observations. What is the critical value for the hypothesis test shown below, assuming alpha = 0.05?

$$H_0: \rho = 0$$

$$H_1: \rho < 0$$

   a. –1.868
   b. –1.717
   c. –1.714
   d. –1.645

6. What is the value of the test statistic for $r = 0.56$ and $n = 10$?
   a. $t = 1.91$
   b. $t = 2.03$
   c. $t = 2.39$
   d. $t = 3.39$

7. What is the interpretation of the slope in a simple linear regression?
   a. If y-hat increases by 1 unit, x will increase by the amount of the slope.
   b. If x increases by 1 unit, y-hat will change by the amount of the slope.
   c. If x equals 0, y-hat will equal the slope.
   d. An increase in x will cause y-hat to increase.

Use the following information to answer questions 8 and 9.

A warehouse manager wants to be able to predict processing time from the number of pallets. They record the time it takes to process orders according to the number of pallets required for eight orders. The data are shown below:

	Pallets	**Time**			
Observation	X	Y	XY	X^2	Y^2
1	30	20	600	900	400
2	45	27	1215	2025	729
3	58	35	2030	3364	1225
4	65	42	2730	4225	1764
5	70	58	4060	4900	3364
6	85	52	4420	7225	2704
7	95	76	7220	9025	5776
8	98	67	6566	9604	4489
SUM	**546**	**377**	**28,841**	**41,268**	**20,451**

8. What is the equation of the estimated regression line?
   a. $\hat{y} = 5.309 + 0.777x$
   b. $\hat{y} = 5.605 + 0.877x$
   c. $\hat{y} = -5.906 + 0.777x$
   d. $\hat{y} = -5.956x + 0.877$

9. What is the value of R^2, the coefficient of determination?

   a. 0.60

   b. 0.70

   c. 0.80

   d. 0.90

10. A simple linear regression model was estimated to predict the number of customer returns based on the number of sales the previous month.

$$\hat{y} = 230 + 0.0034x$$

    What is the expected number of returns if last month's sales equaled 45,678 units?

   a. 155.3

   b. 230

   c. 385.3

   d. 456.8

11. A simple linear regression model returns a negative residual for a particular value of X. What must be true about the model?

   a. The regression model is underestimating the observation.

   b. The regression model is overestimating the observation.

   c. The regression model has a negative slope.

   d. The regression model is overestimating the input variable.

12. What are the assumptions for simple linear regression?

   a. Y is independent, normally distributed with constant mean m and standard deviation sigma.

   b. Residuals are independent and normally distributed with constant mean equal to 0 and variance = σ^2.

   c. X and Y are normally distributed and dependent.

   d. Errors are dependent, normally distributed with mean equal to 0 and variance = σ^2/n.

13. A 95% confidence interval on the response of a regression model will be at its narrowest point when which statement is true?

   a. $X = y$

   b. $X = 0$

c. X = x-bar

d. $X = b_1$

14. For a simple linear regression model, an analyst calculates 95% confidence and prediction intervals. What is true about these intervals?

    a. At any given X, the confidence interval will be wider than the prediction interval.

    b. The confidence and prediction intervals will be equal at X = x-bar.

    c. Both the confidence and prediction intervals will get wider as x deviates from its average value.

    d. The width of the prediction interval is independent of the value of x.

15. A Black Belt estimates a regression model to be

$$\hat{y} = -32.5 + 12.0x_1 + 0.45x_2 - 3.9x_1x_2.$$

    What type of model is this?

    a. Multiple linear regression model

    b. Nonlinear regression model

    c. Simple linear regression model

    d. Logistic regression model

16. The purpose of discriminant analysis is to:

    a. develop a linear combination of input variables that acts to separate the data into two or more classes.

    b. detect small differences in the accuracy of gages.

    c. reduce the dimensionality of a data set when inputs are correlated.

    d. categorize brainstormed ideas into natural groupings.

17. What type of error occurs when the null hypothesis is rejected when H_0 is true?

    a. Alpha

    b. Beta

    c. Gamma

    d. No error; it's a correct decision.

18. Two hypothesis tests were run to test whether a new training program decreased mean call times for two call centers in different cities. The p-value of the test for the first center was equal to 0.04 and the p-value for the second center was equal to 0.001. At $\alpha = 0.05$, what can be said about the mean call times for the centers?

    a. The second call center had a larger decrease in mean call time.

    b. The training program worked better in the first center.

    c. The training program significantly decreased call times in each call center.

    d. The training program had an equal effect on call times for each center.

19. A Black Belt performs a hypothesis test, and the analysis returns a p-value equal to 0.045. What is the interpretation of the p-value?

    a. The null hypothesis is true with a probability equal to 0.045.

    b. The alternative hypothesis is true with a probability equal to 0.045.

    c. The probability of getting the current result or something more extreme assuming the null hypothesis is true is 0.045.

    d. The probability of getting the current result or something more extreme assuming the alternative hypothesis is true is equal to 0.045.

20. Which of the following pairs of hypotheses is appropriate to test the difference between two means?

    a. H_0: $\mu_d = 0$ vs. H_1: $\mu_d \neq 0$

    b. H_0: $\mu_1 - \mu_2 = 0$ vs. H_1: $\mu_1 - \mu_2 \leq 0$

    c. H_0: $\mu_1 - \mu_2 = 0$ vs. H_1: $\mu_1 - \mu_2 \neq 0$

    d. H_0: $\bar{x}_1 - \bar{x}_2 = 0$ vs. H_1: $\bar{x}_1 - \bar{x}_2 \leq 0$

21. A training company has shown that its study program results in a significantly higher mean score on a 100-point certification exam compared to those who did not use the program. Further investigation shows that the mean score for those who used the training program was 80.4, compared to 80.2 for those who did not. This is an example of:

    a. a false claim.

    b. a result being statistically insignificant.

    c. a result being statistically insignificant but practically significant.

    d. a result being statistically significant, but not practically significant.

22. After making an adjustment to a filling process, a Six Sigma professional wants to determine with 95% confidence whether the mean fill weight changed by 0.5g. The process historical standard deviation is $\sigma = 2.5$ g. What is the minimum required sample size needed to make this determination?

    a. 48

    b. 49

    c. 96

    d. 97

23. A process has an average of 5.50 for the last 20 parts produced and a historical standard deviation of 0.20. Find the 95% prediction interval for the next part produced.

    a. [5.098, 5.902]

    b. [5.108, 5.588]

    c. [5.171, 5.829]

    d. [5.412, 5.588]

24. A process output is normally distributed with a sample mean equal to 6.75 and a sample standard deviation of 0.40. If 100 parts are produced, what is the two-sided tolerance interval that we can be 99% confident will contain 95% of the products produced?

    a. [5.580, 7.920]

    b. [5.678, 7.822]

    c. [5.810, 7.690]

    d. [5.927, 7.573]

25. A sample of $n = 35$ is taken. The sample mean is equal to 2.15, and the sample standard deviation is 0.80. Calculate the 90% confidence interval for the mean.

    a. [1.875, 2.425]

    b. [1.921, 2.379]

    c. [1.925, 2.375]

    d. [1.928, 2.372]

26. Which of the following statements is a correct interpretation of a 95% confidence interval on the mean?

    a. We are 95% confident that the population will fall in the interval.

    b. We are 95% confident that the mean falls in the interval.

    c. We are 95% confident that the interval covers the true mean.

    d. We are 95% confident that the sample mean equals the true mean.

27. Samples were taken from two normal distributions. Construct a 95% two-sided confidence interval on the ratio of the two variances using the following information: $n_1 = 25$, $s_1 = 48.0$, $n_2 = 25$, $s_2 = 38.0$.

    a. [0.679, 3.740]

    b. [0.704, 3.621]

    c. [0.716, 3.557]

    d. [0.805, 3.165]

28. A 90% one-sided confidence interval on the ratio of two standard deviations is calculated to be $[0.657, \infty)$. What can we conclude?

    a. The first standard deviation is significantly larger than the second.

    b. The second standard deviation is significantly larger than the first.

    c. The interval does not include zero. Conclude that the standard deviations are significantly different.

    d. The interval includes one. Conclude that the standard deviations are not significantly different.

29. A hypothesis test will be conducted on the difference between means of two normally distributed groups. It cannot be assumed that the variances of the two groups are equal. What are the degrees of freedom that should be used for the test given $s_1 = 10.0$, $n_1 = 20$, $s_2 = 15.0$, and $n_2 = 15$?

    a. 23

    b. 25

    c. 32

    d. 33

30. To determine the efficacy of a weight loss regimen, participants are weighed at the start and at the end of the 12-week program. Which pair of hypotheses would be appropriate?

   a. H_0: $\mu_1 = \mu_2$ vs. H_1: $\mu_1 \neq \mu_2$

   b. H_0: $\mu_1 - \mu_2 = 0$ vs. H_1: $\mu_1 - \mu_2 > 0$

   c. H_0: $\mu_D = 0$ vs. H_1: $\mu_D \neq 0$

   d. H_0: $\mu_D = 0$ vs. H_1: $\mu_D > 0$

31. To use the chi-square distribution to test a population variance, the data must be independent and:

   a. correlated.

   b. normally distributed.

   c. categorized.

   d. uniformly distributed.

32. A Six Sigma team has put a new guide on a drilling machine to reduce the variation in the location of a drilled hole. Before improvements, the standard deviation of the location was equal to 0.005. Ten parts are run through the milling operation, and the resulting sample standard deviation is equal to 0.003. Test if the change has significantly reduced the standard deviation of the hole location at the 95% level.

   a. Test statistic = 3.24; do not reject H_0.

   b. Test statistic = 3.24; reject H_0.

   c. Test statistic = 2.28; do not reject H_0.

   d. Test statistic = 2.28; reject H_0.

33. Before an improvement project, the percentage defective for a milling process was equal to 5.7%. After the changes were made, 150 parts were run with a percent defective equal to 4.0%. Use the normal approximation method to test at the 95% level if the changes significantly reduced the percentage of defective parts.

   a. $Z_0 = -0.90$; reject H_0.

   b. $Z_0 = -0.90$; do not reject H_0.

   c. $Z_0 = -1.06$; reject H_0.

   d. $Z_0 = -1.06$; do not reject H_0.

Section VI Questions

34. A new electric vehicle is tested for its maximum range under a single charge. Four vehicles were tested on a track, with the following mileage results: {260, 252, 254, 254}. Assuming mileage is normally distributed, test whether the average range is at least 250 miles using an alpha = 0.05.

    a. $z_0 = 2.89$; reject H_0.

    b. $z_0 = 2.89$; do not reject H_0.

    c. $t_0 = 2.89$; reject H_0.

    d. $t_0 = 2.89$; do not reject H_0.

35. To compare the means of three or more groups, which of the following tests is the most appropriate?

    a. Student's *t*-test

    b. Fisher's exact test

    c. Chi-squared test

    d. ANOVA

Use the following ANOVA table to answer questions 36 and 37.

Source of Variation	Sum of Squares	Degrees of Freedom	Mean Square	*F*-statistic
Factor A	144	2		
Factor B				
AB Interaction	360	8		
Error	75			
Total	664	29		

36. Given the two-way ANOVA table, what is the value of the F-statistic for Factor B?

    a. 2.13

    b. 4.25

    c. 9.0

    d. 14.4

37. Based on the ANOVA table, which of the following statements is correct?

    a. Factors A and B are significant at alpha = 0.05.

    b. Factor A and the AB interaction are significant at alpha = 0.05.

    c. Factor B and the AB interaction are significant at alpha = 0.05

    d. Factors A, B, and the AB interaction are significant at alpha = 0.05.

38. A Black Belt performs chi-square goodness-of-fit test with alpha = 0.05 to determine whether the distribution of defect types last month was different from the historical percentages. Given the data table below, which is statement is true?

	Defect Type		
	Length	**Weight**	**Strength**
Historical %	50%	25%	25%
Monthly Inspection	60	35	45

a. The test statistic = 3.889 and the critical value = 5.991.

b. The test statistic = 4.286 and the critical value = 5.991.

c. The test statistic = 3.889 and the critical value = 7.815.

d. The test statistic = 4.286 and the critical value = 7.815.

39. With n observations from a random sample, each of which could be classified into exactly one of k categories, how many degrees of freedom does a goodness-of-fit test with specified probabilities have?

a. n

b. $n - 1$

c. $k - 1$

d. $k - 2$

40. The SWIFT technique can be used to identify:

a. strengths and weaknesses in an organization.

b. root causes of a problem.

c. risks and causes.

d. wastes in a process.

41. Items that are classified as high supply risk, but low profit impact are considered:

a. noncritical items.

b. bottleneck items.

c. strategic items.

d. leverage items.

42. An organization has identified a low-priority risk. What action might they take to address this risk?

a. Tolerate the risk.

b. Terminate the risk.

c. Treat the risk.

d. Transfer the risk.

43. What would be the best classification for a risk that is minor and is likely to occur?

a. Low

b. Medium

c. High

d. Critical

44. An organization outsources a component that requires specialty equipment and highly trained operators. What type of risk action has it taken?

a. Tolerate the risk.

b. Terminate the risk.

c. Treat the risk.

d. Transfer the risk.

45. An organization may use which of the following strategies so that it can be aware of cybersecurity risks?

a. Update and upgrade all software immediately.

b. Segregate networks using application-aware defenses.

c. Transition to multifactor authentication.

d. Actively manage systems and configurations.

46. Events are placed on a risk matrix according to their:

a. impact and likelihood.

b. detection and severity.

c. impact and severity.

d. likelihood and detection.

47. What actions should an organization take for strategic items in a supplier risk matrix?

a. Develop long-term relationships with suppliers.

b. Substitute the products.

c. Concentrate on process efficiency.

d. Target pricing strategies.

48. A process improvement team has completed a PFMEA. Which of the following tools would be best suited to help the team focus their efforts based on the PFMEA?

    a. SWOT analysis

    b. Pareto analysis

    c. Force-field analysis

    d. Statistical analysis

49. A value indicating the relative risk of a potential failure is referred to as:

    a. failure.

    b. modality.

    c. RPN.

    d. severity.

50. An FMEA has a severity of 7, an occurrence rating of 5, and a detection score of 3. What is the RPN?

    a. 12

    b. 15

    c. 35

    d. 105

51. A systematic approach that proactively identifies, analyzes, prioritizes, and documents potential failure modes and their respective potential causes of failures is:

    a. DFSS.

    b. FMEA.

    c. SIPOC.

    d. PDCA.

52. The likelihood that current controls will prevent a failure from reaching the customer is called:

    a. detection.

    b. occurrence.

    c. severity.

    d. FMEA.

Use the following FMEA to answer questions 53–55.

	Current Process				Action Results			
Row	Severity	Occurrence	Detection	RPN	Severity	Occurrence	Detection	RPN
1	3	5	10	150	3	5	4	60
2	5	10	3	150	3	5	3	45
3	10	3	5	150	10	3	1	30

53. On which row of the FMEA should a process improvement team focus their efforts for the current state of the process?
    a. 1
    b. 2
    c. 3
    d. The RPNs are the same; no special effort is necessary.
54. Give the rows of the current state of the process in order of highest to lowest risk.
    a. 2, 3, 1
    b. 3, 1, 2
    c. 3, 2, 1
    d. The RPNs are the same, so no ranking is needed.
55. After action has been taken, which row now shows the highest risk?
    a. Row 1
    b. Row 2
    c. Row 3
    d. None; the risk has been mitigated.
56. What does the following pair of statements exemplify?

    Current state: Product has 95% reliability at 1000 cycles.

    Future state: Product has 99% reliability at 1000 cycles by year end.

    a. Gap analysis
    b. Value stream analysis
    c. Risk analysis
    d. Waste analysis

57. A causal factor that, if removed, will prevent the recurrence of the same situation is known as a/an:

    a. process control metric.

    b. root cause.

    c. root cause analysis.

    d. FMEA.

58. A3 analysis:

    a. uses PDCA at its core.

    b. is a form of affinity analysis.

    c. is a statical test used for grouped data.

    d. runs counter to DMAIC principles.

59. Which tool is based on the principle of "the vital few and the useful many"?

    a. Affinity diagram

    b. Fishbone diagram

    c. Pareto chart

    d. Process decision program chart

60. Which of the following tools can be used to drill down to a root cause?

    a. Eight wastes

    b. 5S

    c. SMED

    d. 5 Whys

61. Which of the following tools is used to provide visual identification of many potential causes of a problem?

    a. Decision tree

    b. Process flowchart

    c. Cause-and-effect diagram

    d. Check sheet

Use the following diagram to answer questions 62 and 63.

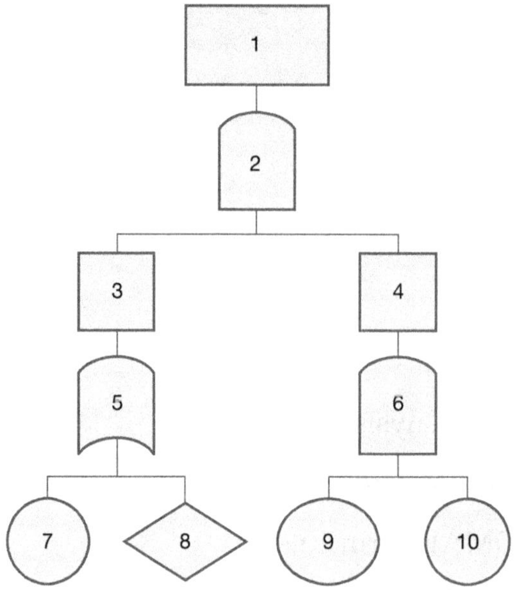

62. Which of the FTA symbols indicates an AND gate?

    a. 2 and 6

    b. 3 and 4

    c. 5

    d. 7, 9, and 10

63. Which of the FTA symbols indicates the primary event?

    a. 1

    b. 3

    c. 4

    d. 3 and 4

64. When constructing a Pareto diagram, where is the "other" category placed?

    a. Left-most column.

    b. Right-most column.

    c. Depends on the number of other items.

    d. The "other" column should not be used.

Use the following paragraph to answer questions 65 and 66.

> ABC Manufacturing, a machine shop that uses total productive maintenance methodologies, received a purchase order from their customer for 50 parts. Because of ABC's experience with manufacturing and the long lead time to produce these parts, ABC creates a production order for 55 parts to ensure that they will complete at least 50 parts.

65. Producing 55 parts when the order was for 50 is an example of which of the Eight Wastes?
    a. Transportation
    b. Time
    c. Defects
    d. Overprocessing

66. ABC Company is actively eliminating what type of waste?
    a. Inventory
    b. Processing
    c. Waiting
    d. Transportation

67. It can be assumed that the most likely reason ABC is producing 55 parts instead of 50 parts is due to what type of waste?
    a. Overprocessing
    b. Defects
    c. Transportation
    d. Inventory

68. In an effort to detect a critical defect, quality inspectors are asked to perform a 100% inspection. This activity is an example of which type of waste?
    a. Time
    b. Defects
    c. Transportation
    d. Overprocessing

SOLUTIONS

1. b; The plotted points decrease from left to right along the y-axis as we move along the x-axis. This is indicative of a negative correlation. The variables X and Y are moving in opposite directions. [VI.A.1]

2. d; The points plotted in Figure D suggest a parabolic, or nonlinear, relationship. [VI.A.1]

3. c; To calculate the sample correlation coefficient from the information given in the table, we note that $n = 8$, and then perform the following steps:

   1. Calculate x-bar.

   $$\bar{x} = \frac{\sum x_i}{n} = \frac{20.0}{8} = 2.5$$

   2. Calculate y-bar.

   $$\bar{y} = \frac{\sum y_i}{n} = \frac{344}{8} = 43$$

   3. Calculate SS_{xy}.

   $$\begin{aligned} SS_{xy} &= \sum x_i y_i - n\bar{x}\bar{y} \\ &= 899.5 - 8(2.5)(43) \\ &= 39.5 \end{aligned}$$

   4. Calculate SS_{xx}.

   $$\begin{aligned} SS_{xx} &= \sum x_i^2 - n\bar{x}^2 \\ &= 59 - 8(2.5^2) \\ &= 9.0 \end{aligned}$$

   5. Calculate SS_{yy}.

   $$\begin{aligned} SS_{yy} &= \sum y_i^2 - n\bar{y}^2 \\ &= 15016 - 8(43^2) \\ &= 224 \end{aligned}$$

   6. Calculate the sample correlation coefficient, r.

   $$\begin{aligned} r &= \frac{SS_{xy}}{\sqrt{SS_{xx}SS_{yy}}} \\ &= \frac{39.5}{\sqrt{9 \times 224}} \\ &= 0.88 \end{aligned}$$

   [VI.A.1]

4. c; To estimate the slope of a simple linear regression line, we can use

$$\begin{aligned} b_1 &= \frac{SS_{xy}}{SS_{xx}} \\ &= \frac{39.5}{9.0} \\ &= 4.39. \end{aligned}$$

Alternatively, we can calculate the slope using the sample correlation coefficient:

$$\begin{aligned} b_1 &= \frac{r\sqrt{SS_{yy}}}{\sqrt{SS_{xx}}} \\ &= \frac{0.88\sqrt{224}}{\sqrt{9}} \\ &= 4.39. \end{aligned}$$

[VI.A.1]

5. b; The critical value for a one-sided, less-than hypothesis test on the correlation is $-t\ (\alpha, n - 2)$. Using a Student's t table, we find $-t(0.05, 22) = -1.717$. [VI.A.1]

6. a; The value of the test statistic for a hypothesis test on the correlation coefficient is equal to

$$\begin{aligned} t &= \frac{r}{\sqrt{1-r^2}}\sqrt{n-2} \\ &= \frac{0.56}{\sqrt{1-0.56^2}}\sqrt{10-2} \\ &= 1.91. \end{aligned}$$

[VI.A.1]

7. b; If x increases by one unit, then the change in the predicted y value is equal to the slope. Proof:

$$\begin{gathered} \frac{\Delta y}{\Delta x} = b_1 \\ \text{Let } \Delta x = 1 \\ \frac{\Delta y}{1} = b_1 \\ \Delta y = b_1 \end{gathered}$$

[VI.A.2]

8. c; To find the estimated slope of the regression line:

$$\bar{x} = \frac{\sum x}{n} = \frac{546}{8} = 68.25$$

$$\bar{y} = \frac{\sum y}{n} = \frac{377}{8} = 47.125$$

$$SS_{xy} = \sum xy - n\bar{x}\bar{y} = 28841 - 8 \times 68.25 \times 47.125 = 3110.75$$

$$SS_{xx} = \sum x^2 - n\bar{x}^2 = 41268 - 8 \times 68.25^2 = 4003.5$$

$$b_1 = \frac{SS_{xy}}{SS_{xx}} = \frac{3110.75}{4003.5} = 0.777.$$

Next, we calculate the y-intercept of the regression line:

$$b_0 = \bar{y} - b_1\bar{x} = 47.125 - 0.777 \times 68.25 = -5.906.$$

[VI.A.2]

9. d; The coefficient of determination for a simple linear regression is the square of the correlation coefficient, r. We can find the correlation coefficient using SS_{yy} as well as SS_{xy} and SS_{xx} that were calculated for question 8.

$$SS_{yy} = \sum y^2 - n\bar{y}^2 = 20451 - 8 \times 47.125^2 = 2684.875$$

$$r = \frac{SS_{xy}}{\sqrt{SS_{xx}SS_{yy}}} = \frac{3110.75}{\sqrt{4003.5 \times 2684.875}} = 0.9488$$

$$R^2 = 0.9488^2 = 0.90$$

[VI.A.2]

10. c; For x = sales = 45,678 units, the expected number of returns is equal to

$$\hat{y} = 230 + 0.0034x = 230 + 0.0034 \times 45{,}678 = 385.3.$$

[VI.A.2]

11. b; The residuals of a regression model are calculated using $e = y - \hat{y}$. If a residual is negative, then the predicted value is greater than the observation y. [VI.A.2]

12. b; The assumptions for simple linear regression are that residuals are independent and normally distributed with constant mean 0 and standard deviation sigma. [VI.A.2]

13. c; Both the confidence and prediction intervals from a regression line will be at their narrowest when X = x-bar. [VI.A.2]

14. c. Confidence and prediction intervals are narrowest at X = x-bar, and they get wider as the value of X deviates from its average. [VI.A.2]

15. a; The equation $\hat{y} = -32.5 + 12.0x_1 + 0.45x_2 - 3.9x_1x_2$ is an example of a multiple linear regression model. It has more than one independent variable predicting the response. [VI.A.3]

16. a; The purpose of discriminant analysis is to develop a linear combination of input variables that acts to separate (discriminate) the data into two or more classes. [VI.A.3]

17. a; If we incorrectly reject the null hypothesis when it is true, this is an error of action, also known as an alpha or Type I error. [VI.B.1]

18. c; Since each p-value is less than $\alpha = 0.05$, we can state that the training program was successful at each center. The value of p addresses the strength of the conclusion but not the magnitude of the effect. [VI.B.1]

19. c; By definition, the $p = 0.045$ is interpreted as the probability of getting the current result of something more extreme assuming the null hypothesis is true. [VI.B.1]

20. c; The hypothesis test for the difference in two means is written as

$$H_0\text{: } \mu_1 - \mu_2 = 0 \text{ vs. } H_1\text{: } \mu_1 - \mu_2 \neq 0.$$

The hypotheses can also be expressed as

$$H_0\text{: } \mu_1 = \mu_2 \text{ vs. } H_1\text{: } \mu_1 \neq \mu_2.$$

[VI.B.1]

21. d; This is an example of a result being statistically significant, but not practically significant. [VI.B.2]

22. d; To find the minimum sample size needed to achieve a margin of error of 0.5 g for a 95% confidence interval on the mean, we use

$$n = \left(\frac{z_{\alpha/2}\sigma}{E}\right)^2 = \left(\frac{1.96 \times 2.5}{0.5}\right)^2 = 96.04 \uparrow 97.$$

Note that we always round up the result. [VI.B.3]

23. a; To calculate a 95% prediction interval with x-bar = 5.5, $n = 20$, and $\sigma = 0.20$, we use

$$\left[\bar{x} - z_{\alpha/2}\sigma\sqrt{1+\frac{1}{n}},\ \bar{x} + z_{\alpha/2}\sigma\sqrt{1+\frac{1}{n}}\right]$$

$$\left[5.5 - 1.96(0.20)\sqrt{1+\frac{1}{20}},\ 5.5 + 1.96(0.20)\sqrt{1+\frac{1}{20}}\right]$$

$$[5.098, 5.902].$$

[VI.B.4]

24. c; To calculate a tolerance interval so that we are 99% confident that the interval will contain 95% of the population, we find the tolerance factor K based on $n = 100$. Then

$$LTL = \overline{x} - Ks = 6.75 - 2.350 \times 0.40 = 5.810$$
$$UTL = \overline{x} + Ks = 6.75 - 2.350 \times 0.40 = 7.690.$$

[VII.B.4]

25. b; To construct a 90% two-sided interval on the mean, we note that x-bar = 2.15 and $n = 35$. We find the Student's t value $t(0.05, 34)$ from the table to be 1.691. Then

$$\overline{x} \pm t_{\alpha/2,n-1}\frac{s}{\sqrt{n}}$$
$$2.15 \pm 1.691\frac{0.80}{\sqrt{35}}$$
$$2.15 \pm 0.228665$$
$$[1.921, 2.379].$$

[VI.B.4]

26. c; The mean doesn't try to fall into an interval; it is the interval that tries to cover the mean. The correct interpretation is "We are 95% confident that the interval covers the true mean." [VI.B.4]

27. b; To construct the two-sided interval on the ratio of two variances, we note that the degrees of freedom for the F-statistic are $v_1 = n_1 - 1 = 24$ and $v_2 = n_2 - 1 = 24$. Then

$$\left[\frac{s_1^2}{s_2^2}F_{1-\alpha/2,v_2,v_1}, \frac{s_1^2}{s_2^2}F_{\alpha/2,v_2,v_1}\right]$$
$$\left[\frac{s_1^2}{s_2^2}F_{0.975,24,24}, \frac{s_1^2}{s_2^2}F_{0.025,24,24,}\right]$$
$$\left[\frac{48^2}{38^2}0.44, \frac{48^2}{38^2}2.27\right]$$
$$[0.704, 3.621].$$

[VI.B.4]

28. d; The interval includes 1; therefore we cannot conclude the standard deviations are significantly different. [VI.B.5]

29. a; To test two means when we cannot assume equal variances, we use Welch's method. The degrees of freedom for the Student's t critical value are calculated using

$$\nu = \frac{\left(\frac{s_1^2}{n_1}+\frac{s_2^2}{n_2}\right)^2}{\frac{\left(s_1^2/n_1\right)^2}{n_1-1}+\frac{\left(s_2^2/n_2\right)^2}{n_2-1}}$$

$$= \frac{\left(\frac{10^2}{20}+\frac{15^2}{15}\right)^2}{\frac{\left(10^2/20\right)^2}{20-1}+\frac{\left(15^2/15\right)^2}{15-1}}$$

$$= \frac{400}{\frac{25}{19}+\frac{225}{14}}$$

$$= 23.0054 \downarrow 23.$$

[VI.B.5]

30. d; Since each participant is weighed two times, this is an application of a paired sample test. To demonstrate that the weight loss program is effective, we would test H_0: $\mu_D = 0$ vs. H_1: $\mu_D > 0$. [VI.B.5]

31. b; To use the chi-square distribution to test a population variance, the data must be independent and normally distributed. [VI.B.5]

32. b; We are testing one standard deviation against a standard. The hypotheses for this test are

$$H_0: \sigma = 0.005$$
$$H_1: \sigma < 0.005.$$

We calculate a chi-square test statistic noting that $n = 10$:

$$\chi_0^2 = \frac{(n-1)s^2}{\sigma^2} = \frac{9 \times 0.003^2}{0.005^2} = 3.24.$$

We compare the test static to the table value $\chi^2_{0.95,9} = 3.325$. If $\chi_0^2 < \chi^2_{0.95,9}$, we will reject H_0. Since 3.24 < 3.325, we reject H_0 and conclude that the standard deviation has decreased. [VI.B.5]

33. b; This is a one-sided hypothesis test on a proportion. The hypotheses are

$$H_0: p = 0.057$$
$$H_1: p < 0.057.$$

We note that n = 150, p_0 = 0.057, and p-hat = 4.0% and use the normal approximation method:

$$z_0 = \frac{(\hat{p}-p_0)}{\sqrt{\frac{p_0(1-p_0)}{n}}} = \frac{(0.04-0.057)}{\sqrt{\frac{0.057(1-0.057)}{150}}} = -0.90.$$

We reject H_0 if $z_0 < -z_{0.05}$. From the table, we find $-z_{0.05} = -1.645$. Since –0.90 is not less than –1.645, we fail to reject H_0. [VI.B.5]

34. We set up the one-sided hypotheses for the mean:

$$H_0: \mu = 250$$
$$H_1: \mu > 250.$$

Because we do not know the population standard deviation, we will use the Student's t distribution for this test.

$$\overline{X} = \frac{\sum x_i}{n} = \frac{260+252+254+254}{4} = 255$$

$$s = \sqrt{\frac{\sum x_i^2 - n\overline{x}^2}{n-1}} = \sqrt{\frac{260^2+252^2+254^2+254^2-4\times 255^2}{4-1}} = 3.46$$

$$t_0 = \frac{254-250}{4.73/\sqrt{4}} = 2.89$$

We will reject H_0 if $t_0 > t(0.05, 3) = 2.353$. Since 2.89 > 2.353, we reject H_0 and conclude that the average mileage is greater than 250. [VI.B.5]

35. d; To compare the means of three or more groups, we use analysis of variance (ANOVA). [VI.B.6]

36. b; The completed ANOVA table is shown below:

Source of Variation	Sum of Squares	Degrees of Freedom	Mean Square	*F*-statistic
Factor A	144	2	72	14.4
Factor B	85	4	21.25	4.25
AB Interaction	360	8	45	9
Error	75	15	5	
Total	**664**	**29**		

where:

$$SS_B = SS_{Total} - SS_A - SS_{AB} - SS_E = 664 - 144 - 360 - 75 = 85$$

$$\begin{aligned} DF_{AB} &= DF_A \times DF_B \\ 8 &= 2 \times DF_B \\ DF_B &= 4 \end{aligned}$$

$$\begin{aligned} DF_{Error} &= DF_{Total} - DF_A - DF_B - DF_{AB} \\ &= 29 - 2 - 4 - 8 \\ &= 15 \end{aligned}$$

$$MS_B = \frac{SS_B}{DF_B} = \frac{85}{4} = 21.25$$

$$MS_E = \frac{SS_E}{DF_E} = \frac{75}{15} = 5$$

$$F_B = \frac{MS_B}{ME_E} = \frac{21.25}{5} = 4.25.$$

[VI.B.6]

37. d; The critical values for the *F* tests are shown below:

Factor A: $F(0.05, 2, 15) = 3.68$. $F_0 > F(0.05, 2, 15)$. Reject H_0, conclude Factor A is significant.

Factor B: $F(0.05, 4, 15) = 3.05$. $F_0 > F(0.05, 2, 5)$. Reject H_0, conclude Factor B is significant.

AB interaction: $F(0.05, 8, 15) = 2.64$. $F_0 > F(0.05, 8, 5)$. Reject H_0, conclude AB interaction is significant.

[VI.B.6]

38. b; Using the table, we note that 140 defects were found last month. We then calculate the expected number of defects using the historical percentages:

	Defect Type		
	Length	Weight	Strength
Historical %	50%	25%	25%
Observed Monthly Inspection	60	35	45
Expected	140 × 0.50 = 70	140 × 0.25 = 35	140 × 0.25 = 35

We then calculate the chi-square test statistic using

$$\chi_0^2 = \frac{\sum_{i=1}^{k}(O_i - E_i)^2}{E_i}$$
$$= \frac{(60-70)^2}{70} + \frac{(35-35)^2}{35} + \frac{(45-35)^2}{35}$$
$$= 4.286.$$

The chi-square critical value has (k – 1) = (3 – 1) = 2 degrees of freedom. Using the chi-square table, we find $\chi^2(0.05, 2) = 5.991$. [VI.B.7]

39. c; For a chi-squared goodness-of-fit test, there will be k – 1 degrees of freedom. [VI.B.7]

40. c; The SWIFT technique can be used to identify risks and causes. [VI.C.1]

41. b; Items that are classified as high supply risk but low profit impact are considered bottleneck items. [VI.C.1]

42. a; An organization has identified a low-priority risk. They may choose to tolerate the risk. [VI.C.1]

43. b; For a risk that is minor and is likely to occur, the best classification of the risk is medium. [VI.C.1]

44. d; An organization that outsources a component that requires specialty equipment and highly trained operators has transferred the risk. [IV.C.1]

45. b; To detect a cybersecurity risk, an organization may segregate networks using application-aware defenses. [IV.C.1]

46. a; Events are placed on a risk matrix according to their impact and likelihood. [VI.C.1]

47. a; An organization should develop long-term relationships with suppliers for strategic items in a supplier risk matrix. [IV.C.1]

48. b; A Pareto analysis would be best suited to help the team focus their efforts based on the PFMEA. [VI.C.2]

49. c; The RPN, or risk priority number, is a value indicating the relative risk of a potential failure. [VI.C.2]

50. d; RPN = $S \times O \times D = 7 \times 5 \times 3 = 105$. [VI.C.2]

51. b; Failure mode and effects analysis (FMEA) is a systematic approach that proactively identifies, analyzes, prioritizes, and documents potential failure modes and their respective potential causes of failures. [VI.C.2]

52. a; The likelihood that current controls will prevent a failure from reaching the customer is called detection. [VI.C.2]

53. c; If there is a tie in RPNs, the team should compare the severity ratings and act on the row with the highest severity score. Here, Row 3 shows a severity = 10. This should be the team's focus. [VI.C.2]

54. c; Since the RPNs are tied, we rank the risks of the rows by severity. Row 3 has the highest severity, followed by Row 2, then Row 1. [VI.C.2]

55. a; The RPN of Row 1 is now the highest risk with RPN = 60. [VI.C.2]

56. a; The pair of statements is an example of gap analysis. [VI.D.1]

57. b; A causal factor that, if removed, will prevent the recurrence of the same situation is known as a root cause. [VI.D.2]

58. a; A3 analysis uses PDCA at its core. [VI.D.2]

59. c; The Pareto chart is based on the principle of the vital few and useful many, also known as the 80/20 rule. [VI.D.2]

60. d; The 5 Whys tool can be used to drill down to a root cause. [VI.D.2]

61. c; A cause-and-effect, or fishbone, diagram is used to provide visual identification of many potential causes of a problem. [VI.D.2]

62. a; Symbols 2 and 6 denote AND gates. The output for an AND gate occurs only if all inputs occur. [VI.D.2]

63. a; Symbol 1 represents the primary, or head, event. This is the failure mode being investigated. [VI.D.2]

64. b; In a Pareto chart, the "Other" column should appear as the last column. [VI.D.2]

65. d; Producing more than needed is an overprocessing waste. [VI.D.3]

66. c; ABC Company is trying to eliminate waiting waste due to long lead times. [VI.D.3]

67. b; We can assume that ABC produces 55 parts to deliver 50 parts due to defect waste. [VI.D.3]

68. d; One hundred percent human inspection is an example of overprocessing waste. [VI.D.3]

Section VII

Improve

VII. Improve (21 Questions)

A. ***Design of experiments (DOE)***

1. *Terminology*. Define basic DOE terms, e.g., independent and dependent variables, factors and levels, response, treatment, error, and nested. (Understand)
2. *Design principles*. Define and apply DOE principles, e.g., power, sample size, balance, repetition, replication, order, efficiency, randomization, blocking, interaction, confounding, and resolution. (Apply)
3. *Planning experiments*. Plan and evaluate DOEs by determining the objective, selecting appropriate factors, responses, and measurement methods, and choosing the appropriate design. (Evaluate)
4. *One-factor experiments*. Understand when to use completely randomized, randomized block, and Latin square designs. (Understand)
5. *Two-level fractional factorial experiments*. Understand these types of experiments and describe how confounding can affect their use. (Understand)
6. *Full-factorial experiments*. Understand these types of experiments. (Understand)

B. ***Lean methods***

1. *Waste elimination*. Select and apply tools and techniques for eliminating or preventing waste, e.g., pull systems, kanban, 5S, standard work, and poka-yoke. (Analyze)
2. *Cycle-time reduction*. Use various tools and techniques for reducing cycle time, e.g., continuous flow, single-minute exchange of die (SMED), and heijunka (production leveling). (Analyze)
3. *Kaizen*. Define and distinguish between kaizen and kaizen blitz and describe when to use each method. (Apply)

4. *Other improvement tools and techniques.* Identify and describe how other process improvement methodologies are used, e.g., theory of constraints (TOC), and overall equipment effectiveness (OEE). (Understand)

C. ***Implementation.*** Develop plans for implementing proposed improvements, including conducting pilot tests or simulations, and evaluate results to select the optimum solution. (Evaluate)

QUESTIONS

1. An experimental treatment is:

   a. a specific combination of factor levels.

   b. an input to an experiment.

   c. the relationship between a factor and a response variable.

   d. an independent variable or assignable cause.

2. How do designed experiments differ from trial-and-error, ad hoc testing methods?

   a. A designed experiment can establish causality by changing inputs under controlled conditions.

   b. The number of samples in a designed experiment is unknown before the study is completed.

   c. Interactions cannot be estimated with a designed experiment.

   d. Ad hoc testing methods guarantee that model parameters can be estimated and statistically tested for significance.

3. Design of experiments (DoE) is most closely related to which Six Sigma concept?

   a. 1.5σ shift

   b. 3.4 dpmo

   c. $Y = f(X)$

   d. OEE

4. In a designed experiment, an effect is:

   a. the level of an experimental factor.

   b. an experimental treatment combination.

   c. the difference between an observation and its estimated value.

   d. the change in the response when the level of one or more factors is changed.

5. Experimental error is defined as:

   a. measurement mistakes made during the experimental run.

   b. errors in setting up the experimental treatments.

   c. variation in the response not accounted for by the experimental factors.

   d. mistakes made during the design phase of an experiment.

6. An engineer is studying factors that are important in the annealing process. Annealing is performed at two different facilities, and each facility has three different ovens. Three batches will be run through each oven. The best experimental design for this situation is a:

   a. mixture design.

   b. general factorial design.

   c. nested design.

   d. screening design.

7. For each run in a designed experiment, four parts are produced and measured. This is an example of:

   a. replication.

   b. repetition.

   c. confounding.

   d. blocking.

8. What experimental design technique would be used when an experiment may have to be conducted over several shifts?

   a. Blocking

   b. Confounding

   c. Replication

   d. Randomization

9. Two or more factors, together, produce a result different than the sum of their separate effects. This is known as:

   a. randomization.

   b. orthogonal effects.

   c. replication.

   d. interaction.

10. When conducting a designed experiment, what is the main difference between repetition and replication?
    a. Repetition reflects the sources of variability both within and between runs.
    b. Replication reflects the sources of variability both within and between runs.
    c. Repetition can only be used for fractional factorial designs.
    d. Replication can only be used for fractional factorial designs.

11. An experimenter is investigating the performance of two injection molding machines. For the first machine, raw material from Supplier A is used to produce plastic switches. The second machine produces the switches using material from Supplier B. The factors machine and supplier are said to be:
    a. confounded.
    b. biased.
    c. repeated.
    d. blocks.

12. What is true about blocking variables in a designed experiment?
    a. They are assumed to interact with each factor.
    b. They are not of direct experimental interest.
    c. They increase the number of experimental runs.
    d. They are indicative of a poorly designed experiment.

13. The main purpose of randomization in a designed experiment is to:
    a. minimize the effect of uncontrolled noise factors.
    b. incorporate variation into the responses.
    c. make running the experiment more difficult.
    d. to account for a factor that adds variability to the response.

Use the following information to answer questions 14 and 15.

A Black Belt conducted a three-factor full factorial experiment, with each factor having two levels.

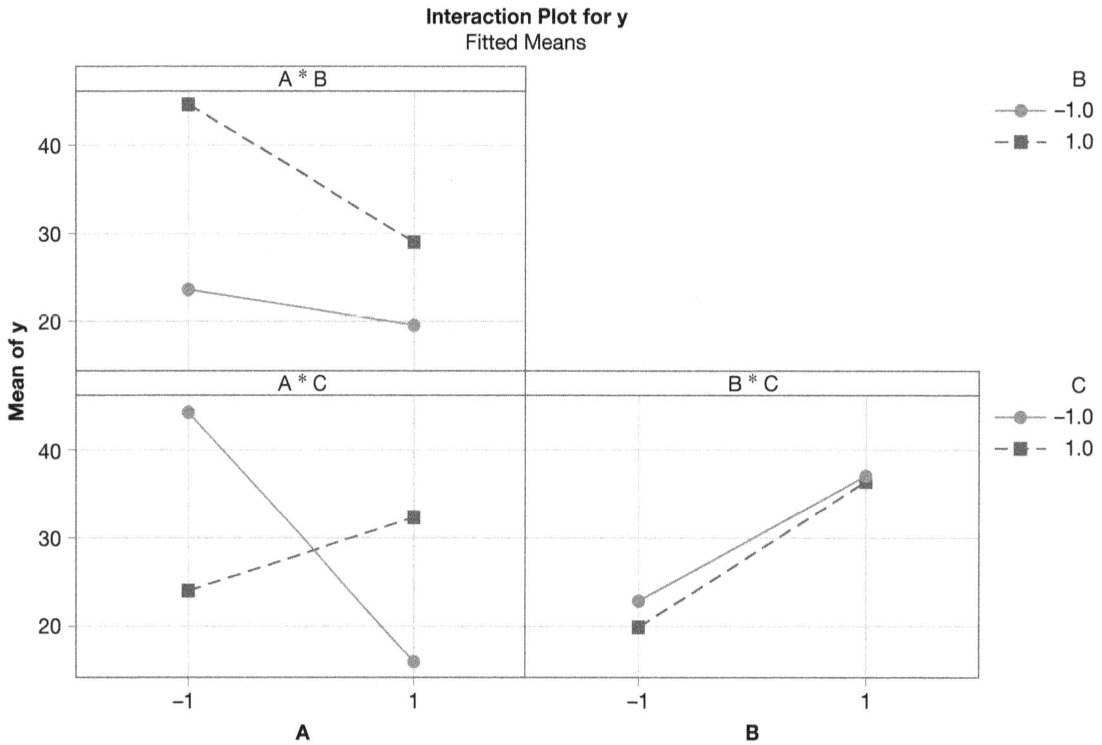

14. The ANOVA table showed that there are two significant interaction effects. Which factors have significant interactions?

    a. AB and BC

    b. AB and AC

    c. BC and AB

    d. BC and AC

15. If Factor A is set at the low level, what is the effect of changing Factor C from low to high on the mean response?

    a. The response does not change.

    b. The response increases.

    c. The response decreases.

    d. The response cannot be determined using the graphs.

16. Resolution of an experiment is a concern for which types of designs?

    a. Full factorial

    b. Fractional factorial

c. Mixture

d. Latin square

17. A new process is being investigated, and the team has identified 12 factors that may affect the response variable of tensile strength. Which experiment should the team plan?

a. Full factorial design

b. Latin square design

c. Screening design

d. One factor at a time design

18. What is the minimum number of runs needed to conduct a full-factorial experiment with five factors, at two levels each?

a. 2

b. 10

c. 25

d. 32

19. Some principles of designed experiments include:

a. one factor at a time and sequential design.

b. all-in-one experiments and multitude of effects.

c. sparsity of effects and sequential design.

d. maximizing confounding effects and heredity of effects.

20. If an experimenter wished to have two factors at two levels each, and one factor at three levels, which type of experimental design would be appropriate?

a. Placket-Burman

b. General factorial

c. Fractional factorial

d. Mixture

21. A randomized block experiment is one in which:

a. each block contains all the treatments.

b. each block is randomly assigned to a treatment.

c. there is no restriction on randomization for blocks or treatments.

d. blocks are treated as confounding effects.

22. A Latin square design is one in which:

    a. there is one control factor and two blocks, all with the same number of levels.

    b. there are two control factors and one block, all with the same number of levels.

    c. blocking variables interact with other blocking variables.

    d. the factor of interest is confounded with at least one block.

23. An experiment is being conducted to optimize a packaging process. One factor at three levels and two blocking factors at three levels each will be used to run the experiment. The experimenter would most likely select what type of design for this experiment?

    a. Latin square

    b. Split plot

    c. Fractional factorial

    d. Mixture

24. In a Resolution III experiment:

    a. main effects are confounded with other main effects.

    b. some main effects are confounded with two-factor interactions.

    c. some two-factor interactions are confounded with the mean.

    d. some main effects are confounded with the mean.

25. A fractional-factorial experiment shows that there is a significant two-factor interaction between factors A and B. However, this effect is confounded with the interaction of factors C and D. What might the experimenter do next?

    a. Run a follow up factorial experiment to de-alias the interactions effects.

    b. Run a one-factor at a time experiment.

    c. Run a Plackett-Burnam design.

    d. Add blocking variables to the design and rerun.

26. In a resolution IV experiment:

    a. main effects are confounded with other main effects.

    b. main effects are confounded with two-factor interactions.

    c. some two-factor interactions are confounded with other two-factor interactions.

    d. main effects are confounded with the mean.

27. How many runs are required for a 2^{4-1} experiment?

    a. 8

    b. 9

    c. 15

    d. 16

28. How many fewer runs would a 2^{6-2} experiment require compared to a 2^6 full-factorial experiment?

    a. 2

    b. 4

    c. 16

    d. 48

29. What design of experiments principle can be used to justify running a fractional factorial experiment instead of a full factorial?

    a. Sparsity of effects

    b. Replication

    c. Randomization

    d. Blocking

30. In a two-level full-factorial design, the effect of Factor A is equal to –12.4. What is the corresponding coefficient for Factor A in the coded regression model?

    a. $-12.4 \div \sqrt{\text{MSE}}$

    b. –6.2

    c. 0

    d. –12.4

31. A 2^2 full-factorial experiment is run with three replicates. The contour plot of continuous Factors A and B against the estimated response displays curved lines. What does this plot indicate?

    a. The model is insignificant.

    b. There is an interaction between factors A and B.

    c. The model needed a blocking factor.

    d. The experimental error is quite large.

32. In a designed experiment, a main effect plot with a steep slope indicates:
    a. significant confounding.
    b. the factor has a large effect on the response.
    c. curvature in the model.
    d. a large effect on the error term.
33. A two-factor interaction graph from a full-factorial experiment displays two lines that cross. This is an indication of:
    a. strong interaction effects between the factors.
    b. an error made in charting.
    c. confounding effects between the factors.
    d. cancelation of the effects of the factors.
34. What is one advantage of adding center points to a factorial experiment?
    a. Allows for a cubic term to be fit in the model
    b. Increases the degrees of freedom of the error term
    c. Allows the experimenter to test whether there is curvature in the model
    d. Increases the resolution of the design
35. When analyzing the results of a 2^k full-factorial experiment, which effects should be examined first in the ANOVA table?
    a. Main effects
    b. Interaction effects
    c. Confounding effects
    d. Resolution effects
36. In a 2^3 full-factorial design, each factor level:
    a. appears in only two experiment runs.
    b. appears in $k = 3$ experimental runs.
    c. is considered a main effect.
    d. is run with every possible configuration of the other factor levels.
37. How many runs will a 2^5 full-factorial experiment with two replicates require?
    a. 50
    b. 64

c. 20

d. 32

38. In a push system, parts move along an assembly process:

    a. as soon as they are completed.

    b. when the customer requests it.

    c. when the next machine is ready.

    d. at the end of the day.

39. A system that simplifies and improves the restocking procedure is called:

    a. 5S.

    b. poka-yoke.

    c. Kanban.

    d. Heijunka.

40. What occurs in the second step of the 5S technique?

    a. Items that do not belong are removed from the workspace.

    b. Necessary items are arranged and organized.

    c. The area is cleaned and shined.

    d. The process is mistake proofed.

41. An error in 5S implementation includes:

    a. cleaning too thoroughly.

    b. thinking of it as a one-time exercise.

    c. failing to mistake-proof.

    d. standardizing organizing and cleaning steps.

42. Work activities that are precisely described and include cycle time, task sequence, and other information is known as:

    a. a pull system.

    b. standard work.

    c. poka-yoke.

    d. kaizen.

43. Standard work describes work activities in terms of:

    a. product specifications.

    b. machine downtime.

    c. human motion.

    d. continuous flow.

44. The shopping tab on a website will not allow the user to continue to the next screen until a correct credit card number is entered. This is an example of:

    a. mistake-proofing.

    b. a bottleneck.

    c. a pull system.

    d. heijunka.

45. Building a new fixture that prevents a metal part from being placed into a drilling machine upside down is an example of:

    a. heijunka.

    b. poka-yoke.

    c. kanban.

    d. theory of constraints.

46. A company forecasts customer demand and sets up a production schedule to meet that demand. This is an example of:

    a. a push system.

    b. a pull system.

    c. heijunka.

    d. one-piece flow.

47. With the SMED technique, emphasis is placed on:

    a. moving internal work to external work.

    b. single-piece flow.

    c. batch processing.

    d. statistical analysis.

48. Two types of heijunka include:

    a. push and pull systems.

    b. transportation and motion.

    c. volume leveling and product leveling.

    d. defect detection and mistake proofing.

49. A customer order for 1000 quarters is due in five days. ABC runs one continuous eight-hour shift per day. What is the takt time?

    a. 0.48 min/piece

    b. 2.4 min/piece

    c. 25 piece/min

    d. 625 pieces/shift

50. A customer order for 150,000 phone cases is due in three days. The factory runs one eight-hour shift per day. What lean tool will help with reducing the impact of maintenance on the process?

    a. Poka-yoke

    b. SMED

    c. 5S

    d. Standard work

51. Heijunka is a technique that addresses:

    a. volume leveling and continuous improvement.

    b. production leveling and theory of constraints.

    c. production leveling and volume leveling.

    d. Kaizen and eight wastes.

52. The main difference between kaizen and a kaizen blitz is that:

    a. kaizen is used for short-term fixes.

    b. a kaizen blitz rarely is an effective way to improve a process.

    c. kaizen is focused on a specific area of production.

    d. a kaizen blitz has a schedule and a deadline.

53. The five steps in the theory of constraints are, in order:

    a. identify, subordinate, elevate, exploit, repeat.

    b. identify, elevate, exploit, subordinate, repeat.

    c. identify, subordinate, exploit, elevate, repeat.

    d. identify, exploit, subordinate, elevate, repeat.

Use the following scenario to answer questions 54–57.

> An injection molding process runs one eight-hour shift per day with 30 minutes for lunch and 15 minutes for planned maintenance. The line was down for 48 minutes due to a broken fitting and was idle for an additional 14 minutes waiting for material. Planned production per shift is 10,000 units. At the end of the shift, 7000 good parts and 195 bad parts were produced.

54. What is the calculated availability?

    a. 0.4206

    b. 0.8575

    c. 0.8708

    d. 0.9609

55. What is the value of the performance?

    a. 0.7000

    b. 0.7195

    c. 0.9355

    d. 0.9450

56. What is the value of the quality metric?

    a. 0.7000

    b. 0.8575

    c. 0.9721

    d. 0.9729

57. Which of the following actions would be best in order to increase the overall equipment effectiveness (OEE)?

    a. Increase the availability of the process.

    b. Increase the quality of the process.

    c. Decrease the number of days before the product is due.

    d. Increase the performance of the process.

58. An evaluation method for a set of solutions that involves manufacturing a smaller production run of production-equivalent parts is known as a:

    a. simulation.

    b. pilot run.

    c. prototype.

    d. model.

59. An evaluation method for a set of solutions that involves using software to evaluate options is a:

    a. pilot run.

    b. prototype.

    c. model.

    d. simulation.

60. An evaluation method for a set of solutions that uses a smaller 3D representation of the production part is a:

    a. pilot run.

    b. simulation.

    c. physical model.

    d. prototype.

Use the following prioritization matrix to answer questions 61 and 62. The potential solutions are ranked with 1 being the best.

	Implementation Cost	**Outside Resources Needed**	**Time to Implement**	**Positive Impact on Customer**
Relative Importance (%)	20%	15%	30%	35%
Potential Solutions				
A	2.5	1	3	3
B	2.5	3	1.5	2
C	1	4	1.5	4
D	4	2	4	1

61. Which criterion is considered most important by the team?
    a. Implementation cost
    b. Outside resources needed
    c. Time to implement
    d. Positive impact on the customer
62. Which potential solution should be considered first?
    a. Solution A
    b. Solution B
    c. Solution C
    d. Solution D
63. Instructions on how to inform the company about an improved process would most likely be found in the:
    a. reaction plan.
    b. communication plan.
    c. mission statement.
    d. quality objectives.

SOLUTIONS

1. a; An experimental treatment is a specific combination of factor levels. [VII.A.1]
2. a; In contrast to trial-and-error methods, a designed experiment can establish causality by changing inputs under controlled conditions. [VII.A.1]
3. c; Design of experiments is most closely aligned with the Six Sigma concept of $Y = f(X)$. [VII.A.1]
4. d; An effect is the change in the response when the level of one or more factors is changed. [VII.A.1]
5. c; Experimental error is defined as the variation in the response not accounted for by the experimental factors. [VII.A.1]
6. c; The best experimental design for this situation is a nested design because each facility has its own different ovens. [VII.A.1]
7. b; For each treatment combination in a designed experiment, four parts are produced and measured. This is an application of repeated measures, or repetition. [VII.A.2]
8. a; If an experiment may have to be conducted over several shifts, it is best to incorporate shifts as blocking variables. The blocks will account for any added variation introduced into the results due to the different conditions during the shifts. [VII.A.2]
9. d; Interaction occurs when two or more factors, together, produce a result different than the sum of their separate effects. [VII.A.2]
10. b; Replication reflects the sources of variability both within and between runs. [VII.A.2]
11. a; The factors machine and supplier are said to be confounded. Their separate effects on the response cannot be estimated. [VII.A.2]
12. b; Blocking variables are not of experimental interest. They are incorporated into an experiment to account for any excess variability introduced by the blocks. [VII.A.2]
13. a; The main purpose of randomization in a designed experiment is to minimize the effect of uncontrolled noise factors. [VII.A.2]
14. b; The lines in the interaction plots for *AB* and *AC* are not parallel, indicating that these factors interact. [VII.A.2]
15. c; From the AC interaction graph, we see that if A is at the low level, moving C from low to high decreases the response. [VII.A.2]

16. b; Resolution is a concern for fractional factorial designs. [VII.A.2]

17. c; To reduce the set of candidate factors as efficiently as possible, the team should plan to run a screening design. [VII.A.3]

18. d; The minimum number of runs for a full factorial experiment with five factors at two levels each is calculated by $2^5 = 32$. [VII.A.3]

19. c; Some principles of designed experiments include the sparsity of effects and sequential design. [VII.A.3]

20. b; A general factorial would be appropriate for a design with two factors at two levels each, and one factor at three levels. [VII.A.3]

21. a; A randomized block experiment is one in which each block contains all the treatments. [VII.A.4]

22. a; A Latin square design is one in which there is one control factor and two blocks, all with the same number of levels. [VII.A.4]

23. a; An example of an application of a Latin square design is having one control factor at three levels and two blocking factors at three levels each. This is a 3×3 Latin square design with nine runs. [VII.A.4]

24. b; In a Resolution III experiment, some main effects are confounded with two-factor interactions. [VII.A.5]

25. a; To de-alias confounded interactions in the fractional-factorial design, the experimenter can run a follow-up experiment (such as a fold-over design). [VII.A.5]

26. c; In a Resolution IV experiment, some two-factor interactions are confounded with other two-factor interactions. [VII.A.5]

27. a; A 2^{4-1} experiment is a half-fraction of a 2^4 experiment. The fractional factorial will require $2^4 \div 2 = 8$ runs. [VII.A.5]

28. d; The full factorial requires $2^6 = 64$ runs. The quarter-fraction $2^{6-2} = 2^4 = 16$ runs. The difference is $64 - 16 = 48$ fewer runs. [VII.A.5]

29. a; The principle of sparsity of effects can be used to justify running a fractional factorial experiment instead of a full factorial. [VII.A.5]

30. b; The coded factor coefficient is equal to one-half of the factor effect. [VII.A.6]

31. b; Curved lines on a contour plot indicate that there is an interaction between factors. [VII.A.6]

32. b; In a designed experiment, a main effect plot with a steep slope indicates that the factor has a large effect on the response. [VII.A.6]

33. a; If a two-factor interaction graph from a full-factorial experiment displays two lines that cross, it is an indication of a strong interaction effect between the factors. [VII.A.6]

34. c; One advantage of adding center points to a factorial design is that it allows the experimenter to test whether there is curvature (a squared term) in the model. [VII.A.6]

35. b; Interaction effects should be examined first when analyzing a 2^k factorial experiment. [VII.A.6]

36. d; In a 2^3 full-factorial design, each factor level is run with every possible configuration of the other factor levels. [VII.A.6]

37. b; The number of runs will be $2^5 \times 2 = 64$. [VII.A.6]

38. a; In a push system, parts move along an assembly process as soon as they are completed. [VII.B.1]

39. c; Kanban is a system that simplifies and improves the restocking procedure. [VII.B.1]

40. b; In the second step of 5S, necessary items are arranged and organized. This is known as Set in Order. [VII.B.1]

41. b; An error in 5S implementation includes thinking of it as a one-time exercise. [VII.B.1]

42. b; Work activities that are precisely described and include cycle time, task sequence, and other information are known as standard work. [VII.B.1]

43. c; Standard work describes work activities in terms of human motion. Each work activity is precisely described, including specifying cycle time, takt time, task sequence, and minimum inventory needed. [VII.B.1]

44. a; The shopping tab on a website will not allow the user to continue to the next screen until a correct credit card number is entered. This is an example of mistake-proofing. [VII.B.1]

45. b; Building a new fixture that prevents a metal part from being placed into a drilling machine upside down is an example of poka-yoke, or mistake-proofing. [VII.B.1]

46. c; A company forecasts customer demand and sets up a production schedule to meet that demand. This is an example of heijunka. [VII.B.2]

47. a; With the SMED technique, emphasis is placed on moving internal work to external work. [VII.B.2]

48. c; Two types of heijunka include volume leveling and product leveling. [VII.B.2]

49. b. Takt time is calculated as the amount of available time divided by the customer demand. [VII.B.2]

$$Takt\,time = \frac{Available\ time}{Customer\,demand} = \frac{5\text{ days} \times 8\text{ hr/day} \times 60\text{ min/hr}}{1000} = \frac{2400}{1000} = 2.4\text{ min/piece}$$

50. c; The SMED, or quick-change-over approach, moves internal work to external work. Internal work is performed during the production run whereas external work is performed before or after the run. Maintenance performed during production will shut down the line. Moving maintenance to before or after the run will help reduce its impact on the process. [VII.B.2]

51. c; Heijunka is a technique for meeting customer demand while reducing variation in production. It has two forms, production and volume leveling. [VII.B.2]

52. d; Kaizen is a term that means gradual unending improvement. A kaizen blitz is a small, intensive team project occurring over a few days to a week. [VII.B.3]

53. d; The five steps of Goldratt's theory of constraints are, in order:

    1. Identify the bottleneck, or constraining process.
    2. Exploit the process by using kaizen or other methods to improve throughput on the constraining process.
    3. Subordinate, or adjust, the rates of the other processes in the chain to match that of the constraining process.
    4. Elevate, or extensively revise, the constraining process to improve the throughput rate.
    5. Repeat the steps to find the new constraint. [VII.B.4]

54. b; The planned operating time is 8 × 60 = 480 minutes, less 30 minutes for lunch and 15 minutes planned maintenance. The actual run time was less due to (48 + 14) = 62 minutes of downtime. [VII.B.4]

$$\begin{aligned} Availability &= \frac{Actual\ operating\ time}{Planned\ production\ time} \\ &= \frac{(8 \times 60) - 30 - 15 - 48 - 14}{(8 \times 60) - 30 - 15} \\ &= \frac{373}{485} = 0.8575 \end{aligned}$$

55. c; To calculate performance, we use

$$Performance = \frac{Total\ pieces / Actual\ operating\ time}{Ideal\ run\ rate}$$
$$= \frac{(7000+195)/(8\times 60-30-15-48-14)}{10{,}0000/(8\times 60-30-15)}$$
$$= \frac{7195/373}{10{,}000/485} = 0.9355.$$

[VII.B.4]

56. d; To calculate quality, we use

$$Quality = \frac{Good\ pieces}{Total\ pieces}$$
$$= \frac{7000}{(7000+195)} = 0.9729.$$

[VII.B.4]

57. a; OEE = Availability × Performance × Quality. Of the three metrics, availability is the lowest. The team should increase the availability of the process to increase throughput. [VII.B.4]

58. b; A pilot run is an evaluation method for a set of solutions that involves manufacturing a smaller production run of production-equivalent parts. [VII.C]

59. d; A simulation is an evaluation method for a set of solutions that involves using software to test the performance of the options. [VII.C]

60. c; A physical model is an evaluation method for a set of solutions that uses a smaller 3D representation of the production part. [VII.C]

61. d; Positive impact on customers has the highest relative importance at 35%. [VII.C]

62. b; Potential solution B has the best overall ranking.

	Implementation Cost	Outside Resources Needed	Time to Implement	Positive Impact on Customer	Total
Relative Importance (%)	20%	15%	30%	35%	100%
Potential Solutions					
A	2.5	1	3	3	2.6
B	2.5	3	1.5	2	2.1
C	1	4	1.5	4	2.65
D	4	2	4	1	2.65

$$\text{Solution A} = (2.5 \times 0.20) + (1 \times 0.15) + (3 \times 0.30) + (3 \times 0.35) = 2.6$$
$$\text{Solution B} = (2.5 \times 0.20) + (3 \times 0.15) + (1.5 \times 0.30) + (2 \times 0.35) = 2.1$$
$$\text{Solution C} = (1 \times 0.20) + (4 \times 0.15) + (1.5 \times 0.30) + (4 \times 0.35) = 2.65$$
$$\text{Solution D} = (4 \times 0.20) + (2 \times 0.15) + (4 \times 0.30) + (1 \times 0.35) = 2.65$$

[VII.C]

63. b; Instructions on how to inform the company about an improved process would most likely be found in the communication plan. [VII.C]

Section VIII

Control

VIII. Control (17 Questions)

A. ***Statistical process control (SPC)***

1. *Objectives.* Explain the objectives of SPC, including monitoring and controlling process performance, tracking trends, runs, and reducing variation within a process. (Understand)
2. *Selection of variables.* Identify and select critical process characteristics for control chart monitoring. (Apply)
3. *Rational subgrouping.* Define and apply the principle of rational subgrouping. (Apply)
4. *Control chart selection.* Select and use control charts in various situations: $\bar{X} - R$, $\bar{X} - s$, individual and moving range (ImR), p, np, c, u, short-run SPC, and moving average. (Apply)
5. *Control chart analysis.* Interpret control charts and distinguish between common and special causes using rules for determining statistical control. (Analyze)

B. ***Other controls***

1. *Total productive maintenance (TPM).* Define the elements of TPM and describe how it can be used to consistently control the improved process. (Understand)
2. *Visual controls.* Define the elements of visual controls (e.g., pictures of correct procedures, color-coded components, and indicator lights), and describe how they can help control the improved process. (Understand)

C. ***Maintain controls***

1. *Measurement system reanalysis.* Review and evaluate measurement system capability as process capability improves and ensure that measurement capability is sufficient for its intended use. (Evaluate)

2. *Control plan*. Develop a control plan to maintain the improved process performance, enable continuous improvement, and transfer responsibility from the project team to the process owner. (Apply)

D. ***Sustain improvements***

1. *Lessons learned*. Document the lessons learned and benefits realized from all phases of a project and identify strategies for reinforcing and replicating improvements. (Apply)
2. *Documentation*. Develop or modify documents including standard operating procedures (SOPs), work instructions, and control plans to ensure that the improvements are sustained over time. (Apply)
3. *Training for process owners and staff*. Develop and implement training plans that are handed off to process owners to ensure consistent execution of revised process methods, KPIs to confirm sustained benefits, and standards to maintain process requirements. (Apply)
4. *Ongoing evaluation*. Identify and apply tools (e.g., control charts and control plans) for ongoing evaluation of the improved process, including monitoring leading indicators, lagging indicators, and additional opportunities for improvement. (Apply)

QUESTIONS

1. Common cause variation is considered to be the noise of the process. This type of variation is:
   a. highly controllable.
   b. risk based.
   c. the operator's responsibility.
   d. management's responsibility.
2. Control charts are graphical tools used to monitor a process. Control charts generally do not contain or use:
   a. a centerline.
   b. control limits.
   c. specification limits
   d. subgroups.
3. One of the purposes of statistical process control is to:
   a. map cause and effect.
   b. determine when to take action on a process.
   c. set specification limits to reasonable values.
   d. provide strategic metrics.
4. A primary feature of a statistical process control chart is that:
   a. it differentiates between common and special causes of variation.
   b. over time, it can lead to overadjustments to the process.
   c. it signals when common cause variation is present.
   d. it can be used only for continuous data.
5. An X-bar and *R* chart signals an out-of-control condition. What can be said about the process?
   a. The process output is out of specification.
   b. The distribution of the process output has changed.
   c. Common cause variation is acting on the process.
   d. The process has too many defects.

6. The area between the upper and lower control limits of a control chart is often referred to as:

   a. the voice of the process.

   b. the special cause variation zone.

   c. the specification zone.

   d. the voice of the customer.

7. The process variable to be selected for control charting is sometimes determined by:

   a. which variable is performing the best.

   b. which variable has the least impact on the output.

   c. which variable is a lagging indicator.

   d. contractual requirements from the customer.

8. The idea behind rational subgroups is that:

   a. subgroup units should be as similar as possible.

   b. maximizing subgroup variation improves detection.

   c. subgroups should represent common and special causes of variation.

   d. subgroups represent the "between" variation in a control chart.

9. Pieces from five spot-welding stations flow onto a common conveyor belt. An operator chooses five consecutive pieces each hour to test the tensile strength of the welds. How might this procedure change once an X-bar *R* chart is used to track the tensile strength?

   a. Rational subgrouping will have to be considered.

   b. The tensile strength must be converted to pass/fail data.

   c. The specifications for tensile strength will have to be adjusted.

   d. No changes will be needed.

10. A control chart will be created to monitor a process that has an automated vision system to measure height, the critical feature. A subgroup size of 15 will be used. Which of the following control charts is best suited for this application?

    a. *X-mR*

    b. X-bar and *S*

    c. *np*

    d. EWMA

11. A powder coating line will be monitored for quality. The number of defects per item will be counted using a constant sample size. Which of the following control charts is best suited for this application?

    a. *p*

    b. *np*

    c. *X-mR*

    d. *c*

12. Which of the following subgroup sizes for an $\overline{X} - R$ chart would most likely be used if measurement were costly?

    a. 1

    b. 2

    c. 3

    d. 4

13. An *R* chart is used rather than an *S* chart because it is:

    a. easier to hand-calculate.

    b. more sensitive to variation.

    c. more accurate.

    d. useful for large subgroup sizes.

14. If only one measurement can be taken at a time, which chart can be used to monitor the process?

    a. *p*

    b. Median

    c. EWMA

    d. X-bar and *R*

15. Which chart is a user-friendly control chart for the operator?

    a. Moving average

    b. Median

    c. *S* chart

    d. EWMA

16. An S chart might be used as a measure of dispersion rather than an R chart because it is:

    a. easier to understand.

    b. easier to create.

    c. a better measure of variation.

    d. more sensitive to variations.

17. Why is the range chart not recommended for sample size greater than 9?

    a. The range is too difficult to calculate with large sample sizes.

    b. The range is not robust against outliers.

    c. The range is too small when sample size increased.

    d. The range is unbiased.

18. Calculate the upper control limit of an X-bar chart using the following information:

$$\bar{\bar{X}} = 90.475,\ n = 6,\ \bar{R} = 12.440$$

    a. 84.4665

    b. 96.4835

    c. 97.6529

    d. 108.2270

19. Calculate the lower control limit of an R chart using the following information:

$$\bar{\bar{X}} = 345.50,\ n = 7,\ \bar{R} = 16.25$$

    a. 0

    b. 1.235

    c. 9.441

    d. 12.350

Use the following information to answer questions 20 and 21.

$$\bar{\bar{X}} = 29.87,\ \bar{s} = 3.55,\ n = 16$$

20. Calculate the upper and lower control limits for the S chart.

    a. [1.232, 5.868]

    b. [1.289, 5.811]

c. [1.590, 5.510]

d. [1.519, 5.581]

21. Calculate the upper and lower control limits for the X-bar chart.

a. [27.16, 32.58]

b. [29.11, 29.93]

c. [29.19, 30.55]

d. [29.70, 30.04]

Use the following information to answer questions 22 and 23.

A single sample was measured each hour over the course of 25 hours. An individuals and moving range with period $n = 2$ will be constructed using the following summary statistics:

$$\sum_{i=1}^{k} X_i = 674.50, \sum_{i=1}^{k} mR_i = 24.40$$

22. What are the control limits of the individuals chart?

a. [0, 36.29]

b. [24.05, 29.91]

c. [24.27, 29.69]

d. [26.20, 27.76]

23. What are the control limits of the moving range chart?

a. [0, 2.71]

b. [0, 3.32]

c. [0.47, 1.57]

d. [0.57, 1.46]

e. [0.57, 1.46]

24. A p-chart is used for:

a. monitoring the count of nonconformities in a product.

b. monitoring the fraction nonconforming.

c. cases in which there are few opportunities for nonconformities to occur.

d. continuous data.

25. A fraction nonconforming chart is frequently converted to a percent nonconforming chart:

    a. so that employees can understand it better.

    b. because of the difficulty of variable part-to-part size.

    c. so that an average subgroup size (n) can be used.

    d. so that the lower control limit will be eliminated.

26. What is one purpose of a *p*-chart?

    a. To determine the average quality level

    b. To discover how many defects are in each product

    c. To track defects per unit

    d. To track the between sample variability of measurements

27. The burden of calculating control limits for a *p*-chart with differing subgroup sizes can be minimized by:

    a. changing the product.

    b. using the average subgroup size.

    c. using a *c*-chart.

    d. measuring fewer items

28. What action should be taken if the lower limit of a *c*-chart is negative?

    a. Plot as calculated.

    b. Set the lower limit to zero.

    c. Recalculate; an error has been made.

    d. Use a *u*-chart.

29. A *u*-chart measures the:

    a. count of nonconformities.

    b. rate of defectives per sample.

    c. count of defectives per unit.

    d. rate of nonconformities per unit.

30. A *p*-chart would be best to use in which of the following situations?

    a. The number of surface defects in lots of 10 stainless steel panels will be tracked.

    b. The number of nonconforming light bulbs in a constant sample size of 300 bulbs will be tracked.

    c. The fraction of defective printed circuit boards produced each shift will be tracked. The number of boards produced per shift can vary from 400 to 500, depending on what type of chip is being manufactured.

    d. The average number of flaws per computer produced will be tracked. Random samples of 20–30 computers are checked each hour.

31. A *c* control chart would be best to use in which of the following situations?

    a. The number of surface defects found on stainless steel panels will be tracked. Sample sizes are constant.

    b. The number of nonconforming light bulbs in a constant sample size of 300 bulbs will be tracked.

    c. The fraction of defective printed circuit boards produced each shift will be tracked. The number of boards produced per shift can vary from 400 to 500, depending on what type of chip is being manufactured.

    d. The average number of flaws per computer produced will be tracked. Random samples of 20–30 computers are checked each hour.

32. An *np*-chart is used for:

    a. controlling the number of nonconformities in a product.

    b. controlling the fraction nonconforming in a lot.

    c. cases where there are few opportunities for a nonconformity to occur.

    d. controlling the number of nonconforming products.

33. A *c*-chart tracks the:

    a. number of nonconforming units.

    b. rate of defects per unit.

    c. count of nonconformities.

    d. the variation within subgroups.

34. Trial control limits are generally calculated based on at least:

    a. subgroup sizes of 5.

    b. 25 subgroups.

    c. 100 samples.

    d. three days of production.

35. What is a difference between an attributes control chart and a variables control chart?

    a. Variables charts are more informative.

    b. Attributes charts provide more detailed data.

    c. Attributes charts use physical part measurements.

    d. Variables charts track count data.

36. How are the control limits for an attributes chart calculated?

    a. The centerline value plus or minus the mean.

    b. The centerline value plus or minus the range.

    c. The centerline value plus or minus three times the standard deviation.

    d. The centerline value plus or minus a percentage of the average value.

Use the following information to answer questions 37 and 38.

$$\sum np = 55, \sum n = 2800, k = 30$$

37. What is the centerline of the p-chart?

    a. 0

    b. 0.0007

    c. 0.0196

    d. 0.5893

38. Calculate the control limits of the p-chart using the average sample size.

    a. [0, 0.0086]

    b. [0, 0.0627]

    c. [0.0118, 0.0275]

    d. [0.4540, 0.6570]

Use the following information to answer questions 39 and 40.

$$\sum np = 78, k = 25, n = 200$$

39. What is the centerline of the *np*-chart?

  a. 0.016

  b. 0.390

  c. 3.12025

  d. 5.791

40. What are the upper and lower control limits of the *np*-chart?

  a. [0, 1.30]

  b. [0, 7.26]

  c. [0, 8.38]

  d. [2.46, 3.78]

41. Which distribution is used to calculate the control limits of the *np*-chart?

  a. Binomial

  b. Poisson

  c. Geometric

  d. Normal

42. Each hour, 10 units are sampled and inspected for defects. After 25 time periods, the total number of defects equals 38. Using this information, calculate the control limits for a *c*-chart.

  a. [0, 1.32]

  b. [0, 2.69]

  c. [0, 5.22]

  d. [0, 9.65]

43. Which distribution is used to calculate the control limits of a *c*-chart?

  a. Normal

  b. Poisson

  c. Uniform

  d. Student's *t*

Use the following information to answer questions 44 and 45.

$$\sum_{i=1}^{k} c_i = 240, \sum_{i=1}^{k} n_i = 1550, k = 25$$

44. What is the centerline of the u-chart?

    a. 0.006

    b. 0.154

    c. 3.85

    d. 9.60

45. What are the upper and lower control limits for a point with $n = 20$ samples?

    a. [0, 0.059]

    b. [0.131, 0.177]

    c. [0, 0.417]

    d. [0.1241, 0.1838]

46. A run on a control chart refers to:

    a. six consecutive increasing or decreasing points.

    b. a point out of control.

    c. three consecutive points above or below the control limits.

    d. a change in control limits.

47. Inconsistent criteria for defective items (that is, different inspectors have different ideas of what a defective is) can lead to:

    a. periodic shifts on an attributes chart.

    b. a random pattern on an attributes chart.

    c. variable control limits on an attributes chart.

    d. negative lower control limits.

48. Applying all eight Shewhart rules to determine out-of-control conditions on a variables control chart can result in:

    a. a decrease in process variability.

    b. a decrease in sensitivity.

    c. an increase in Type I errors.

    d. an increase in Type II errors.

49. An X-bar chart has nine points in a row above the center line. What is the probability of this occurring, assuming the process is in control?

    a. 0.00195

    b. 0.00270

    c. 0.00450

    d. 0.50000

50. A control chart shows 14 points in a row alternate up and down. What might be a possible cause for this out-of-control condition?

    a. Tool wear

    b. A mean shift

    c. Stratification

    d. Excessive tampering

51. Total productive maintenance combines:

    a. lean principles with push systems.

    b. kanban with the Six Big Losses.

    c. continuous improvement and quality function deployment.

    d. maintenance procedures with continuous improvement initiatives.

52. Which of the following situations is an example of periodic-based maintenance?

    a. A timing belt that has a high probability of failure at 70,000 miles might be scheduled for replacement at 60,000 miles.

    b. A bearing that usually fails between 1800 and 2600 hours, causing a major shutdown, might be replaced on the weekend after it has had 1500 hours of service.

    c. Spot welding tips must be dressed or changed more frequently when used on galvanized steel.

    d. It is necessary to perform an oil change on an automobile every 5000 miles.

53. Which of the following best describes the goal of total productive maintenance (TPM)?

    a. To perform maintenance in a timely manner

    b. To reduce waste from the production line

c. To ensure that there are no unplanned delays in production

d. To ensure that proper production planning has occurred prior to the issuing of material

54. A timing belt on a truck engine has failure time between 60,000 to 80,000 miles. The company has decided to replace the belts at 50,000 miles. This is an example of:

    a. condition-based maintenance.

    b. periodic maintenance.

    c. reliability-based maintenance.

    d. maintenance analytics.

55. A continuous improvement project team began using SPC to track the diameter of a drilled hole in a part. After several tool change cycles, the continuous improvement team noticed that the SPC chart consistently went out of control at part number 190. Changing the preventive maintenance cycle from 200 parts to being triggered when the SPC chart goes out of control would be an example of the use of:

    a. maintenance analytics.

    b. visual controls.

    c. reliability-based maintenance.

    d. process capability.

56. During pollen season, air filters must be changed more often. This is an example of what type of maintenance?

    a. Condition-based

    b. Preventive

    c. Reliability-based

    d. Corrective

57. The refill line on a lubricant container is an example of:

    a. maintenance analytics.

    b. reliability-based maintenance.

    c. process capability.

    d. visual controls.

58. A kaizen event was performed on a process, reducing the process variability from 95% to 60%. The next step the continuous improvement team should implement will be to:

    a. perform another kaizen event.

    b. perform a gage R&R study.

    c. buy a more precise measuring device.

    d. reduce machine downtime.

59. The primary purpose of the control plan is to:

    a. implement validations on specific processes.

    b. ensure that process changes are maintained over time.

    c. to create structure for how quality will be implemented throughout the process.

    d. keep only one item as a variable and maintain all other characteristics as controls per experiment.

60. A control plan lists action steps to take if the critical parameter goes out of control. These actions would most likely be found in which column of the control plan?

    a. Reaction plan

    b. Characteristic

    c. Specification

    d. Frequency

61. The requirements for the critical parameter being controlled would be located in which section of a control plan?

    a. Specification

    b. Characteristic

    c. Reaction plan

    d. Frequency

62. Developing documents such as SOPs, work instructions, and control plans helps ensure that:

    a. the paperwork burden remains high.

    b. internal auditors are kept busy.

c. improvements are sustained over time.

d. the right people get credit for the improvement.

63. Which of the following tools is used to monitor a process to make sure that the process improvements made are continuing to produce the desired results?

a. SOPs

b. Control charts

c. Gage R&R

d. Control plan

64. Standard operating procedures (SOPs) and work instructions can:

a. reduce process variation.

b. allow for gradual improvement over time.

c. aid in benchmarking efforts.

d. create an atmosphere of mistrust.

65. When creating documentation, best practices suggest using:

a. the outline method.

b. graphics and illustrations.

c. statistical terminology.

d. a narrative approach.

SOLUTIONS

1. d; Common cause variation, the noise of the process, is considered to be management's responsibility. [VIII.A.1]
2. c; Control charts generally do not contain or use specification limits. [VII.A.1]
3. b; One of the purposes of statistical process control is to determine when to take action on a process. [VII.A.1]
4. a; A primary feature of a statistical process control chart is that it differentiates between common and special causes of variation. [VIII.A.1]
5. b; An X-bar and *R* chart signals an out-of-control condition, which indicates that the distribution of the process output has changed. [VIII.A.1]
6. a; The area between the upper and lower control limits of a control chart is often referred to as the voice of the process. [VIII.A.1]
7. d; The process variable to be selected for control charting is sometimes determined by contractual requirements from the customer. [VIII.A.2]
8. a; The idea behind rational subgroups is that subgroups units should be as similar as possible. [VIII.A.3]
9. a; Rational sub-grouping will have to be considered. Now, samples are a mixture of pieces from five different stations. [VIII.A.3]
10. b; Use an X-bar and *S* chart when measuring continuous variables with sample sizes greater than or equal to 10. [VIII.A.4]
11. d; Use a *c*-chart when counting defects per sampling unit with constant sample size. [VIII.A.4]
12. b; If measurement were costly, it is likely that a sample size of 2 would be used for the X-bar and *R* chart. A sample size equal to 1 must be plotted using an *X-mR* chart. [VIII.A.4]
13. a; An *R* chart is used rather than an *S* chart because it is easier to hand-calculate. [VIII.A.4]
14. c; An exponentially weighted moving average (EWMA) chart can be used for sample sizes equal to 1. [VIII.A.4]
15. b; A median chart is user-friendly for the operator because it does not require any calculation. An odd sample size is used, and all sample values are plotted on the chart. The middle point, or median, is then circled. [VIII.A.4]
16. c; The standard deviation plotted in an *S* chart is a better measure of variation than the range. The range ignores the data between the extremes. [VIII.A.4]

17. b; Because the range uses the extremes of the sample in its calculation, it is very sensitive to outliers. When the sample size is greater than 9, the probability of an outlier increases. In addition, the range uses only two data values in its calculation, regardless of the sample size. As sample size increases, more and more information is lost. [VIII.A.4]

18. b; To calculate the upper control limit of the X-bar chart, we need to look up the table value A_2, which is based on the subgroup size, n. Then we use

$$UCL_{\bar{X}} = \bar{\bar{X}} + A_2\bar{R}$$
$$A_2 \text{ for } n = 6 \text{ is } 0.483$$

$$UCL_{\bar{X}} = 90.475 + 0.483 \times 12.440 = 96.4835.$$

[VIII.A.4]

19. b; To calculate the lower control limit for the R chart, we need to look up the table value D_3, which is based on subgroup size, n. Then we use

$$LCL_R = D_3\bar{R}$$
$$D_3 \text{ for } n = 7 \text{ is } 0.076$$

$$LCL_R = 0.076 \times 16.25 = 1.235.$$

[VIII.A.4]

20. c; To calculate the upper and lower control limits of an S chart, we need to look up two table values, B_3 and B_4, both based on the subgroup size, n. Then we use

$$LCL_S = B_3\bar{S}$$
$$B_3 \text{ for } n = 16 \text{ is } 0.448$$

$$LCL_S = 0.448 \times 3.55 = 1.590$$

$$UCL_S = B_4\bar{S}$$
$$B_4 \text{ for } n = 16 \text{ is } 1.552$$

$$UCL_S = 1.552 \times 3.55 = 5.510.$$

[VIII.A.4]

21. a; To calculate the control limits of the X-bar chart with $n = 16$, we find the table value $A_3 = 0.763$ based on n. Then

$$LCL = \bar{\bar{X}} - A_3\bar{s} = 29.87 - 0.763 \times 3.55 = 27.16$$
$$LCL = \bar{\bar{X}} + A_3\bar{s} = 29.87 + 0.763 \times 3.55 = 32.58.$$

[VIII.A.4]

22. c; We note that number of samples is $k = 25$, and the period for the moving average is $n = 2$. We first calculate

$$\overline{X} = \frac{\sum_{i=1}^{k} X_i}{k} = \frac{674.5}{25} = 29.98$$

$$\overline{mR} = \frac{\sum_{i=1}^{k} mR_i}{k-1} = \frac{24.40}{24} = 1.017.$$

We find the table value E_2 for $n = 2$ to be 2.660. Then

$$LCL = \overline{X} - E_2\overline{mR} = 26.98 - 2.660 \times 1.017 = 24.27$$
$$UCL = \overline{X} + E_2\overline{mR} = 26.98 + 2.660 \times 1.017 = 29.69.$$

[VIII.A.4]

23. b; To find the control limits for the moving range chart with period $n = 2$, we look up table values $D_3 = 0$ and $D_4 = 3.267$ based on n. Then

$$LCL = D_3\overline{mR} = 0 \times 1.017 = 0$$
$$UCL = D_4\overline{mR} = 3.267 \times 1.017 = 3.32.$$

[VIII.A.4]

24. b; A *p*-chart is used for monitoring the fraction nonconforming. [VIII.A.4]

25. a; A fraction nonconforming chart is frequently converted to a percent nonconforming chart so that employees can understand it better. Percentages are often easier for people to evaluate than decimals or fractions. [VIII.A.4]

26. a; One purpose of a *p*-chart is to determine the average quality level of a product. [VIII.A.4]

27. b; The calculation burden arising from variable subgroup size for a *p*-chart can be minimized by using the average subgroup size. [VIII.A.4]

28. b; If the lower limit of a *c*-chart is negative, set the lower limit to zero. [VIII.A.4]

29. d; A *u*-chart measures the rate of nonconformities (defects) per unit. [VIII.A.4]

30. c; The *p*-chart would be best for tracking the fraction of defective printed circuit boards produced each shift in which there is a varying sample size. [VIII.A.4]

31. a; The *c*-chart would be best to track the number of surface defects on stainless steel panels in lots of constant sample size. [VIII.A.4]

32. d; An *np*-chart is used for controlling the number of nonconforming products. The *np*-chart uses a constant sample size. [VIII.A.4]

33. c; A *c*-chart tracks the count of nonconformities. Sample sizes are constant. [VIII.A.4]

34. b; Trial control limits are generally calculated based on at least 25 subgroups. [VIII.A.4]

35. a; Variables charts that use continuous data are more informative than attributes control charts, which are based on discrete data. [VIII.A.4]

36. c; Control limits for attributes charts are calculated using the centerline plus or minus three times the standard deviation. [VIII.A.4]

37. c; To find the average proportion defective, we use

$$\bar{p} = \frac{\sum_{i=1}^{k} D_i}{\sum_{i=1}^{k} n_i} = \frac{\sum_{i=1}^{k} n_i p_i}{\sum_{i=1}^{k} n_i} = \frac{55}{2800} = 0.0196.$$

[VIII.A.4]

38. b; To find the average *n*, we use

$$\bar{n} = \frac{\sum_{i=1}^{k} n_i}{k} = \frac{2800}{30} = 93.3.$$

Then we can calculate the control limits of the *p*-chart:

$$LCL = \bar{p} - 3\sqrt{\frac{\bar{p}(1-\bar{p})}{\bar{n}}} = 0.0196 - 3\sqrt{\frac{0.0196(1-0.0196)}{93.3}} = -0.0234 \uparrow 0$$

$$LCL = \bar{p} + 3\sqrt{\frac{\bar{p}(1-\bar{p})}{\bar{n}}} = 0.0196 + 3\sqrt{\frac{0.0196(1-0.0196)}{93.3}} = 0.0627.$$

[VIII.A.4]

39. c; To find the centerline of the *np*-chart, we use

$$\overline{np} = \frac{\sum_{i=1}^{k} (np)_i}{k} = \frac{78}{25} = 3.12.$$

[VIII.A.4]

40. c; To calculate the control limits of the *np*-chart, we first must find *p*-bar:

$$\overline{p} = \frac{\sum_{i=1}^{k}(np)_i}{kn} = \frac{78}{25 \times 200} = 0.0156.$$

Then

$$LCL = \overline{np} - 3\sqrt{\overline{np} \times (1-\overline{p})} = 3.12 - 3\sqrt{3.12(1-0.0156)} = -2.138 \uparrow 0$$
$$LCL = \overline{np} + 3\sqrt{\overline{np} \times (1-\overline{p})} = 3.12 + 3\sqrt{3.12(1-0.0156)} = 8.378.$$

[VIII.A.4]

41. a; The *np*-chart is based on the binomial distribution. There is a constant number of independent trials, there are only two possible classifications for each unit in a sample, and the probability of a defective is constant from trial to trial. [VIII.A.4]

42. c; To calculate the control limits for a *c*-chart, we first calculate the average number of defects found each time period, noting that data was collected over a total of $k = 25$ time periods:

$$\overline{c} = \frac{\sum_{i=1}^{k} c}{k} = \frac{38}{25} = 1.52.$$

Then

$$LCL = \overline{c} - 3\sqrt{\overline{c}} = 1.52 - 3\sqrt{1.52} = -2.18 \uparrow 0$$
$$UCL = \overline{c} + 3\sqrt{\overline{c}} = 1.52 + 3\sqrt{1.52} = 5.22.$$

[VIII.A.4]

43. b; The *c*-chart is based on the Poisson distribution. There is a constant number of samples taken over time, and we are counting the number of defects in the sample. [VIII.A.4]

44. b; To calculate the centerline for the *u*-chart, we use

$$\overline{u} = \frac{\sum_{i=1}^{k} c_i}{\sum_{i=1}^{k} n_i} = \frac{240}{1550} = 0.1548.$$

[VIII.A.4]

The variable *k* is the number of subgroups used for the control limits. In this example, there were 240 defects found in 25 subgroups in which a total of 1550 units were inspected.

45. c; To calculate the control limits for the u-chart, we note that u-bar = 0.154 from the previous question, and n_i = 20. Then we use

$$\bar{u} = 0.154 \text{ and } n_i = 20$$

$$LCL = \bar{u} - 3\sqrt{\frac{\bar{u}}{n_i}} = 0.154 - 3\sqrt{\frac{0.154}{20}} = -0.109 \uparrow 0$$

$$LCL = \bar{u} + 3\sqrt{\frac{\bar{u}}{n_i}} = 0.154 + 3\sqrt{\frac{0.154}{20}} = 0.417.$$

[VIII.A.4]

46. a; A run on a control chart refers to six consecutive points that are increasing or decreasing. [VIII.A.5]

47. a; Inconsistent criteria for defective items (that is, different inspectors have different ideas of what a defective is) can lead to periodic shifts on an attributes chart. [VIII.A.5]

48. c; Applying all eight Shewhart rules to determine out-of-control conditions on a chart can result in an increase in Type I errors, also known as false alarms. [VIII.A.5]

49. a; For a symmetric control chart, the probability of a point plotting on one side of the centerline is 0.50. If nine points in a row plot on the same side of the centerline, the probability of occurrence is $(0.50)^9 = 0.00195$. Since this probability is so small, we consider this an out-of-control condition. [VIII.A.5]

50. d; Having 14 points in a row alternating up and down could be caused by excessive tampering by operators who adjust the machine when a point is above the centerline, and then again when a point is below the centerline, even though the points are in control. [VIII.A.5]

51. d; Total productive maintenance combines maintenance procedures with continuous improvement initiatives. [VIII.B.1]

52. d; Performing an oil change every 5000 miles in an automobile is an example of periodic, preventive maintenance. [VIII.B.1]

53. c; TPM is a method that works to ensure that every machine in a process is able to perform its required tasks so that production is never interrupted. TPM maximizes equipment effectiveness by using a preventive maintenance program throughout the life of the equipment. [VII.B.1]

54. c; Changing the timing belt before its mean time to failure is an example of reliability-based maintenance. [VIII.B.1]

55. a; Using information from the SPC chart to determine when the part should be changed is an example of using maintenance analytics. Maintenance analytics is defined as scheduling maintenance tasks based on data from sensing devices. [VIII.B.1]

56. a; Changing air filters more often during pollen season is an example of condition-based maintenance. [VIII.B.1]

57. d; The refill line on the lubricant container is an example of visual controls. [VIII.B.2]

58. b; The next step the team should take after reducing the process variability would be to perform a gage R&R study to determine if the current measurement system is adequate. [VIII.C.2]

59. b; The primary purpose of a control plan is to ensure that process changes are maintained over time. [VIII.C.2]

60. a; Action steps to take when a critical parameter goes out of control would appear in the reaction plan column of a control plan. [VIII.C.2]

61. a; The specification section of the control plan would provide the requirements for parameters being monitored. [VIII.C.2]

62. c; Developing documents such as SOPs, work instructions, and control plans helps assure that improvements are sustained over time. [VIII.C.2]

63. b; Control charts are used to monitor a process to make sure that improvements continue to produce the desired results. [VIII.C.2]

64. a; SOPs and work instructions help ensure that procedures are performed the same way each time, which reduces process variation. [VIII.D.2]

65. b; For creating documentation, best practices suggest using graphics and illustrations. [VIII.D.2]

Section IX

Design for Six Sigma (DFSS) Framework and Methodologies

IX. Design for Six Sigma (DFSS) Framework and Methodologies (6 Questions)

A. ***Common DFSS methodologies.*** Identify and describe DMADV (define, measure, analyze, design, and validate) and DMADOV (define, measure, analyze, design, optimize, and validate). (Understand)

B. ***Design for X (DFX).*** Describe design constraints, including Design for cost, Design for manufacturability (producibility), Design for test, and Design for maintainability. (Understand)

C. ***Robust designs.*** Describe the elements of robust product design, tolerance design, and statistical tolerancing. (Understand)

QUESTIONS

1. What are the phases of a design for Six Sigma (DFSS) project?
   a. Define, measure, analyze, improve, control
   b. Design, define, measure, analyze, validate
   c. Define, measure, analyze, design, verify
   d. Design, define, validate, and verify
2. Design for Six Sigma (DFSS) is a methodology:
   a. to eliminate waste.
   b. to perform kaizen events.
   c. to improve the quality of existing processes to a 6-sigma level.
   d. to design parts that result in a 6-sigma level of quality.
3. Which phase of DMAIC is not present in DFSS methodologies?
   a. Control
   b. Measure
   c. Improve
   d. Analyze
4. The measure phase in DMEDI is different than DMAIC since:
   a. current performance can be measured in DMEDI.
   b. current performance cannot be measured in DMEDI.
   c. DMAIC uses tools such as quality function deployment.
   d. DMAIC has no baseline to evaluate current performance.
5. The V in DMADV stands for a word that means to:
   a. ensure quality is maintained throughout all phases.
   b. ensure that the design output can meet its intended use.
   c. ensure that the design input can meet its intended use.
   d. control the change made throughout the phases.

6. Which of the following features is a similarity between DMAIC and DMEDI?
   a. Both have a define phase.
   b. Both use benchmarking to measure current performance.
   c. Both have an explore phase.
   d. Both use simulations to verify design requirements.
7. The X in Design for X is:
   a. an industry term for Six Sigma.
   b. a variable that can be any constraint to the design process.
   c. an industry term for manufacturing.
   d. a Japanese symbol for zero defects.
8. A major benefit of Design for Cost is:
   a. the ability to reduce costs before production and processes are set.
   b. to determine make versus buy decisions.
   c. justification to increase budget.
   d. that costs after production are not considered.
9. The most important contributor to the success of Design for producibility, manufacturability, and assembly is:
   a. considering costs in the design process.
   b. having assembly or manufacturing subject matter experts in the design process.
   c. having gates for each phase.
   d. management support.
10. Design for test is unique to other Design for X methodologies in the sense that:
    a. parts using design for test are designed to be sampled.
    b. design for test is only used on printed wire boards.
    c. design for test uses one final functional test to verify that a quality part was made.
    d. additional features are added to the design to allow for testing of the product throughout the assembly process.

11. A functional requirement:

    a. defines how a part should be built.

    b. defines the performance of a product under specific conditions.

    c. is a requirement that can only be observed but not verified.

    d. is a requirement that requires validation studies to be performed to verify its effectiveness.

Use the following table to answer questions 12 and 13.

A	B	C
1.0 ± 0.5	0.8 ± 0.2	3.0 ± 0.3

12. Calculate the total tolerance of the overall length using the conventional tolerance method.

    a. 0.62

    b. 1.0

    c. 2.0

    d. 6.8

13. Calculate the total tolerance of the overall length using the statistical tolerance method.

    a. 0.62

    b. 1.0

    c. 2.0

    d. 6.8

14. Taguchi's loss function:

    a. combines cost, target, and variation into one metric.

    b. can only be used for robust designs.

    c. is only concerned with production losses in dollars due to waste of raw material.

    d. is a step-function with steps occurring at the specification limits.

15. What is the most common loss function used for symmetric two-sided tolerances?

    a. Variance is best.

    b. Larger is better.

c. Smaller is better.

d. Nominal is best.

16. Designing around the use of inferior-grade, low-cost components and raw materials in the design process is referred to as:

a. tolerance design.

b. parameter design.

c. Taguchi array.

d. system design.

17. The process of selectively upgrading components to eliminate excessive variation is known as:

a. tolerance design.

b. parameter design.

c. Taguchi arrays.

d. system design.

18. Using the statistical tolerancing method, calculate the lower and upper specifications of the stacked assembly shown below.

	Part A	**Part B**	**Part C**
LSL	0.750	0.355	1.230
USL	0.795	0.360	1.280

a. [2.285, 2.485]

b. [2.318, 2452]

c. [2.335, 2.385]

d. [2.351, 2.419]

19. A 15-piece sample from a normal distribution yields a sample mean equal to 34.56 and a sample standard deviation equal to 2.34. Calculate the two-sided tolerance interval in which we will be 90% confident that the interval will cover 95% of the population.

a. [34.266, 39.348]

b. [34.511, 39.921]

c. [34.700, 40.363]

d. [34.926, 40.892]

20. The principle of Taguchi's loss function is that:

    a. parts outside the control limits are scrapped.

    b. any part not at the nominal value incurs loss.

    c. loss is best modeled by a step function.

    d. loss can be reduced by tightening specifications.

SOLUTIONS

1. c; The phases of a DFSS project include define, measure, analyze, design, and verify (DMADV). [IX.A]
2. d; DFSS is a methodology to design parts that result in a 6-sigma level of quality. [IX.A]
3. a; The control phase is not used in DFSS. [IX.A]
4. d; The measure phase in DMEDI (define, measure, explore, develop, implement) is different than DMAIC since current performance cannot be measured in DMEDI. [IX.A]
5. b; The V in DMADV stands for verify, which ensures that the design output can meet its intended use. This phase is performed on the final product or process. [IX.A]
6. a; Both DMAIC and DMEDI have a common define phase. [IX.A]
7. b; The X in Design for X is a variable that can be any constraint to the design process. [IX.B]
8. a; A major benefit of Design for cost is the ability to reduce costs before production and processes are set. It is difficult to make any substantial cost reductions after the design phase. [IX.B]
9. b; The most important contributor to the success of Design for producibility, manufacturability, and assembly is having assembly or manufacturing subject matter experts in the design process. Other strategies include:
    1. minimize part count.
    2. design parts so they can be used on either side.
    3. use off-the-shelf parts, standard components, and purchased parts.
    4. use modular design where possible. [IX.B]
10. d; Design for test adds features to the design to allow for testing of the product throughout the assembly processes, instead of relying only on a functional test at the end of assembly. [IX.B]
11. b; A functional requirement defines the performance of a product under specific conditions. [IX.B]
12. b; Convention tolerancing is the method used to determine the tolerance when several parts are stacked together. To calculate the tolerance of the overall length, we sum the tolerance of each component:

    $$0.5 + 0.2 + 0.3 = 1.0.$$

    [IX.C]

13. a; Statistical tolerances are used if the processes producing the parts are capable and the dimensions are normally distributed. The tolerances in this case are equal to $3\sigma_A, 3\sigma_B$, and $3\sigma_C$, respectively. Unlike conventional tolerances, standard deviations are not additive. Variances, however, are additive. The total statistical tolerance then can be calculated using

$$\begin{aligned} T_{Stack} &= 3\sqrt{(\sigma_A)^2 + (\sigma_B)^2 + (\sigma_C)^2} \\ &= \sqrt{(3\sigma_A)^2 + (3\sigma_B)^2 + (3\sigma_C)^2} \\ &= \sqrt{T_A^2 + T_B^2 + T_C^2} \\ &= \sqrt{0.5^2 + 0.2^2 + 0.3^2} = 0.62. \end{aligned}$$

[IX.C]

14. a; Taguchi's loss function combines cost, target, and variation into one metric. [IX.C]

15. d; The nominal-is-best loss function is most commonly used for two-sided symmetric tolerances. [IX.C]

16. b; Parameter design is used to make a product or process more robust against variation from inferior-grade, low-cost components, and raw materials. [IX.C]

17. a; Tolerance design is the process of selectively upgrading components to eliminate excessive variation. [IX.C]

18. d; The specifications for Part A are 0.7725 ± 0.0225. For Part B, 0.3575 ± 0.0025 and for Part C, 1.255 ± 0.025. The statistical tolerance can be found by using

$$\begin{aligned} T_{Stack} &= 3\sqrt{(\sigma_A)^2 + (\sigma_B)^2 + (\sigma_C)^2} \\ &= \sqrt{(3\sigma_A)^2 + (3\sigma_B)^2 + (3\sigma_C)^2} \\ &= \sqrt{T_A^2 + T_B^2 + T_C^2} \\ &= \sqrt{0.0225^2 + 0.0025^2 + 0.025^2} = 0.034. \end{aligned}$$

The overall nominal value is equal to

$$Nominal_{Total} = (0.75 + 0.0225) + (0.355 + 0.0025) + (1.23 + 0.025) = 2.385.$$

We then calculate the lower and upper tolerance limits:

$$[2.385 - 0.034,\ 2.385 + 0.034] = [2.351,\ 2.419].$$

[IX.C]

19. d; In order to find the tolerance interval, we must use a table value such as the one found in Appendix J.2 of *The ASQ Certified Six Sigma Handbook*. We look up the K value for 90% confidence, 0.95 proportion, and n = 15 and find K = 2.706.

    Then we use

$$\bar{X} \pm Ks = 34.56 \pm 2.706 \times 2.34$$
$$= [34.926, 40.892].$$

    [IX.C]

20. b; The principle of Taguchi's loss function is that any part not at the nominal value incurs loss. [IX.C]

Part B
Simulated Exam

Part B of the study guide presents 150 questions corresponding to each section of the Black Belt BoK, in random order. The number of questions for each section of the BoK is shown in the table below.

BoK Section	Topic	Number of Questions on Certification Exam
Section I	Organization-Wide Planning and Deployment	12
Section II	Organizational Process Management and Measures	10
Section III	Team Management	18
Section IV	Define	20
Section V	Measure	25
Section VI	Analyze	22
Section VII	Improve	21
Section VIII	Control	15
Section IX	Design for Six Sigma (DFSS) Framework and Methodologies	7

QUESTIONS

1. Some hallmarks of Six Sigma include the use of teams to work on well-defined projects, and an emphasis on:

   a. eliminating waste.

   b. grassroots support.

   c. opinion-led decisions.

   d. a five-step problem-solving process.

2. A system that is designed to implement a process or set of processes is known as a:

   a. quality system.

   b. business system.

   c. Six Sigma system.

   d. delivery system.

3. An effective control plan contains which elements?

   a. Quality, risk, and reaction plans

   b. Specifications, detection, and risk plans

   c. Gage R&R, diagnosis, and risk plans

   d. Monitoring, detecting, diagnosing, and reaction plans

4. What is the difference between learning and performance?

   a. Learning creates potential; performance is the realization of the potential.

   b. Learning creates behavior change; performance creates potential.

   c. Learning is the realization of potential; performance occurs in the classroom.

   d. Learning is the realization of training; performance shows training gaps.

5. A normally distributed process output has an average of 1.50 and a sample standard deviation of 0.20. If 50 parts are produced, what is the tolerance interval that will cover at least 99% of the distribution with 95% confidence?

   a. [0.877, 2.123]

   b. [0.929, 2.071]

   c. [0.986, 2.014]

   d. [1.046, 1.954]

6. Using the conventional method, calculate the total tolerance of the assembly shown below.

Part A	Part B	Part C
1.20 ± 0.05	2.50 ± 0.10	2.0 ± 0.50

a. 0.20

b. 0.65

c. 1.30

d. 5.70

7. What might occur when a consultative decision-making approach is used?

a. The team is asked to give specific information, and then the leader makes the final decision.

b. The team is informed of the situation and each member's opinion is asked for individually. The group is not brought together. The leader then makes the final decision.

c. The leader works with the team to reach group consensus.

d. The leader uses information they already have and makes the decision without input from the team.

8. An organization wishes to provide training to improve employee morale. The first step in developing a training program is to:

a. design a training plan.

b. develop the curriculum.

c. assess the need for the training.

d. hire a certified trainer.

9. A maintenance technician reviews the preventive maintenance checklists and notices that the lubricant level of a CNC mill machine is below the refill line. The technician then fills the tank with more lubricant. Refilling the lubricant is an example of:

a. periodic-based maintenance.

b. corrective-based maintenance.

c. reliability-based maintenance.

d. condition-based maintenance.

10. Which type of improvement team is least affected by geography?
    a. Cross-directional
    b. Self-directed
    c. Virtual
    d. Informal
11. A balanced designed experiment is one in which:
    a. all responses are equivalent.
    b. the number of runs for each factor level is equal.
    c. there are an even number of factors.
    d. the factors are set to the same levels.
12. Material that has entered the process but has not reached the output or finished stage can be defined as:
    a. a bottleneck.
    b. overprocessing waste.
    c. work in progress (WIP).
    d. work in queue (WIQ).

Use the precedence diagram below to answer questions 13 and 14.

Activity	Predecessor	Time
A	—	10
B	—	5
C	A, B	10
D	—	15
E	D	5
F	E	5
G	C, F	5

13. What activities make up the critical path?
    a. ABCG
    b. ACG
    c. BCG
    d. DEFG

14. What is the latest day that Activity D can begin without affecting the project delivery date?
    a. Day 0
    b. Day 3
    c. Day 5
    d. Day 10
15. The term White Belt is usually associated with an individual who:
    a. retains their regular position but is trained in the tools, methods, and skills necessary to conduct Six Sigma improvement projects.
    b. has a basic understanding of Six Sigma principles.
    c. is typically assigned full time to train, mentor, and lead the strategy for the organization's strategic projects.
    d. has the authority or ability to make changes to the process as required.
16. A Six Sigma project is chartered to speed up loan processing times. The team has decided that it will track two metrics: processing time and error counts. Error counts has likely been added to the project metrics to:
    a. appease the quality department.
    b. set the team up for failure.
    c. avoid sub-optimizing the process.
    d. satisfy the project champion.
17. An employee who might experience negative effects from a Six Sigma project would be considered a:
    a. stakeholder.
    b. subject matter expert.
    c. Yellow Belt.
    d. project champion.
18. A tool used in the design process to minimize or prevent failures and improve safety and quality is known as:
    a. SWOT.
    b. PDCA.
    c. DFSS.
    d. DFMEA.

19. A process decision program chart is a type of:

    a. tree diagram.

    b. prioritization matrix.

    c. scatter plot.

    d. project-scheduling tool.

20. An auditor selects 25 financial reports each month and classifies them as complete or incomplete. Which control chart would be appropriate to monitor this process?

    a. *u*-chart

    b. *np*-chart

    c. X-bar *S* chart

    d. *c*-chart

21. The statistical standard for quality associated with Six Sigma is:

    a. 80% of problems are due to 20% of the causes.

    b. 3.4 defects per million opportunities.

    c. yield equal to 99.73%.

    d. RTY = 100%.

22. Which of the following metrics is an example of a tactical performance indicator?

    a. Daily production counts

    b. Market share

    c. Weekly average call length

    d. Control chart measurements

23. The time required to complete one unit from the beginning of the process to the end of the process is defined as:

    a. touch time.

    b. cycle time.

    c. takt time.

    d. throughput.

24. When managing a project, a decrease in project funding can result in:

    a. an increase in time and quality.

    b. an increase in quality and decrease in scope.

    c. a decrease in time and scope.

    d. an increase in quality and decrease in scope.

25. A winery conducts an experiment that varies the proportions of three varietals to create a new blend. The resulting blends are then graded on a tasting scale. The best experimental design for this situation is a:

    a. general factorial design.

    b. nested design.

    c. screening design.

    d. mixture design.

26. The rate in time per unit at which the process must complete units to meet the customer demand is known as:

    a. touch time.

    b. cycle time.

    c. takt time.

    d. set up time.

27. Which of the following items may be a constraint on the effectiveness of a virtual team?

    a. Time zones

    b. Staff schedules

    c. Geography

    d. Travel expenses

28. The C in the SIPOC model refers to:

    a. contractors.

    b. customers.

    c. currency.

    d. capability.

29. A net promoter score:

    a. measures the effectiveness of the sales force.

    b. measures the outcome of an advertising campaign.

    c. can be calculated using a Z score.

    d. measures the likelihood that customers will recommend the brand or product to others.

30. Which lean technique emphasizes the use of items such as guide rails, quick-change fasteners, staging carts, and interchangeable parts?

    a. SMED

    b. JIT

    c. Kanban

    d. Heijunka

31. The team phase in which members understand one another and recognize each other's strengths and weaknesses is known as:

    a. storming.

    b. forming.

    c. performing.

    d. norming.

32. Involving stakeholders in project charter reviews:

    a. helps ensure cooperation.

    b. increases the scope of the project.

    c. moves the team from storming to norming.

    d. increases implementation costs.

33. Categories that are typically evaluated in a Six Sigma maturity model include:

    a. number of employees trained and number of concurrent projects.

    b. opportunity identification and project evaluation.

    c. number of statistical software licenses purchased, and black belts trained.

    d. stock price and reductions in staff.

34. The nodes in an interrelation digraph with the most incoming arrows are considered:

    a. drivers.

    b. outcomes.

    c. contingencies.

    d. correlated.

35. A stakeholder who has been designated as high potential for cooperation and high potential to impact a project would be considered:

    a. a mixed blessing.

    b. supportive.

    c. marginal.

    d. nonsupportive.

36. A typical question a champion may ask in a tollgate review for the analyze phase may be:

    a. What are the critical Xs?

    b. Will we run a pilot?

    c. Who are the stakeholders?

    d. What did the FMEA show?

37. Color-coding, floor markings, and indictor lights are examples of:

    a. kanban.

    b. theory of constraints.

    c. visual management.

    d. 5S.

38. During a tollgate meeting, the project champion concludes that the project is no longer consistent with the goals of the organization. What action should be taken?

    a. More data should be collected.

    b. The team should return to the Improve phase.

    c. The project should be canceled.

    d. The team leader should undergo more training.

39. A company collects performance data only on products that have been in service for at least 1000 hours. The resulting sample is not likely to be representative of the population due to:

    a. exclusion bias.

    b. survivorship bias.

    c. self-selection bias.

    d. recall bias.

40. When addressing disruptive behavior in a meeting, the first choice of the team leader should be to:

    a. make eye contact with the individual.

    b. pull the individual aside during a break.

    c. call out the behavior in front of the team.

    d. walk by while making eye contact.

41. When the occurrence of an event has no influence on the occurrence of a second event, the events are considered to be:

    a. mutually exclusive.

    b. dependent.

    c. multiplicative.

    d. independent.

42. Issues that an asset security continuity plan may address include:

    a. the loss of key personnel.

    b. pandemics.

    c. the loss of intellectual property.

    d. strikes by bargaining unit personnel.

43. When implementing a large project, the tool that should be used to show how the smaller activities are linked to larger activities is called a:

    a. Gantt chart.

    b. interrelationship diagram.

    c. work breakdown structure.

    d. tollgate review.

44. A team constructs a spaghetti diagram to show nurses' movements in the emergency department. There are several intersecting lines in the diagram. This is likely a sign that:

    a. the department layout should be redesigned.

    b. the nurses are not well trained.

    c. the supply cabinet is centrally located.

    d. patients need to be moved through the department faster.

45. A process improvement team has reduced part-to-part variation by 20%. What action should the team consider performing next?

    a. Reevaluating the measurement system

    b. Increasing inventory

    c. Decreasing repeatability

    d. Increasing takt time

46. The team phase in which team members express their own opinions and ideas, often disagreeing with others, is referred to as:

    a. norming.

    b. storming.

    c. forming.

    d. performing.

47. A process requires four steps that operate in series. The first pass yields (FPYs) for the steps are equal to 0.98, 0.945, 0.90, and 0.97, respectively. What is the rolled throughput yield (RTY)?

    a. 0.808

    b. 0.850

    c. 0.945

    d. 0.949

48. Evaluate the adequacy of the following problem statement:

    Customer field complaints for Product X have been too high for the past year.

    a. The problem statement meets all requirements.

    b. The problem statement is missing a measurable item.

c. The problem statement is missing the performance gap and business impact.

d. The problem statement is missing a time component.

49. Which type of data has meaningful differences but no absolute zero?

a. Ratio

b. Ordinal

c. Nominal

d. Interval

50. Multivoting is a technique that is:

a. used to construct an affinity diagram.

b. used as an alternative to an interrelationship digraph.

c. used to select items from a brainstormed list using point assignments.

d. used as a follow up to the critical-to-flow diagram.

51. Which of the following statements concerning lean is true?

a. Lean focuses on reducing variation.

b. Lean focuses on reducing waste.

c. Six Sigma saves money; lean does not.

d. Six Sigma is visual; lean is more analytical.

52. What is the minimum number of runs needed to conduct a full-factorial experiment with three factors, at two levels each?

a. 6

b. 8

c. 9

d. 32

53. Which of the following actions is the last step a company should take when benchmarking?

a. Revise internal performance measurements and goals.

b. Perform gap analysis.

c. Determine whom to benchmark against.

d. Collect internal data.

54. A two-factor interaction graph from a full-factorial experiment displays two lines that are parallel. This is an indication that:

    a. an error was made in charting.

    b. there are confounding effects between the factors.

    c. there is a lack of an interaction effect between the factors.

    d. the effects of the factors are canceling each other.

55. In statistical process control, the aim is to:

    a. maximize within-subgroup variation to minimize between-subgroup variation.

    b. minimize within-subgroup variation to make between-subgroup variation more prominent.

    c. maximize both within and between subgroup variation.

    d. maximize special cause variation.

56. Which type of matrix diagram can be used to show pairwise relationships between three variables, such as customer requirements, internal process metrics, and departments?

    a. L-shaped matrix

    b. T-shaped matrix

    c. X-shaped matrix

    d. Roof-shaped matrix

57. A team leader makes action assignments for each member. The leader is likely trying to get the team to avoid:

    a. the use of groupthink.

    b. the airing of old disputes.

    c. reluctance to participate.

    d. arguing over who gets credit.

58. Work instructions are an important input into:

    a. kanban.

    b. standardized work.

    c. value stream maps.

    d. goal statements.

59. The root cause analysis method that uses AND and OR gates is known as a:

    a. tollgate analysis.

    b. process decision program chart.

    c. failure mode and effects analysis.

    d. fault tree analysis.

60. Key behavior indicators (KBIs) can be used to:

    a. identify loyal customers.

    b. optimize systems.

    c. track operational performance.

    d. track cultural change in an organization.

61. What action should an organization consider for noncritical items in a supplier risk matrix?

    a. Develop long-term relationships with suppliers.

    b. Substitute the products.

    c. Concentrate on process efficiency.

    d. Target pricing strategies.

62. A total of 32,678 units was produced, and 244 units were found to be inoperable. What is the ppm defective?

    a. 746.5

    b. 746.7

    c. 7464.7

    d. 7466.8

63. Which charts often have control limits that vary in width?

    a. X-bar and *R* charts

    b. *p*- and *u*-charts

    c. *c*- and *np*-charts

    d. MA-MR charts

64. A typical question a champion may ask in a tollgate review for the measure phase may include:

    a. What are the critical Xs?

    b. Will we run a pilot?

    c. Who are the stakeholders?

    d. What did the FMEA show?

65. Thirty car hoods are inspected for paint defects. The probability of finding four or more defects can be expressed using:

    a. $\Pr(X = 4)$.

    b. $\Pr(X > 4)$.

    c. $\Pr(X \geq 4)$.

    d. $1 - \Pr(X = 4)$.

66. Takt time can be thought of as:

    a. cycle time with no bottle necks.

    b. the rate of customer pull.

    c. the time to replenish a kanban.

    d. the time it takes to changeover a machine.

67. Ultimately, Six Sigma projects should be consistent with an organization's long-term strategic objectives and:

    a. customer requirements.

    b. stock price.

    c. monthly goals.

    d. current quality levels.

68. An R&D department is planning to create a new product that will be sold through discount retail outlets. The design team will likely use:

    a. Design for Cost.

    b. Design for Manufacturability.

    c. Design for Test.

    d. Design for Service.

69. Defects per unit for a process is found to be 0.40. What is the first pass yield (FPY)?

    a. 0.40

    b. 0.49

    c. 0.60

    d. 0.67

70. A high-performing team has recently experienced turnover in the individuals participating in the team project. With the change in personnel, at what stage is the team now most likely to find itself?

    a. Performing

    b. Adjourning

    c. Norming

    d. Forming

71. What is one advantage of adding center points to a factorial experiment?

    a. Helps detect changes in the experimental environment over time

    b. Allows for a cubic term to be fit in the model

    c. Reduces the degrees of freedom of the error term

    d. Increases the resolution of the design

72. In the Kano model, delighter features are those that will cause:

    a. low customer dissatisfaction if not met, but high customer satisfaction if met.

    b. high customer dissatisfaction if not met, but low customer satisfaction if met.

    c. no customer dissatisfaction if not met, but low customer satisfaction if met.

    d. no customer dissatisfaction if not met, but high customer satisfaction if met.

73. A Black Belt lists the vehicle identification number (VIN) of each car produced on the previous shift in a spreadsheet. They then create a macro that selects 10 VINs for inspection. What type of sampling are they using?

    a. Stratified sampling

    b. Convenience sampling

    c. Simple random sampling

    d. Systematic sampling

74. A circuit board production line uses continuous process monitoring to track the hourly totals of defective items. Which of the following control charts is best suited for this application?

    a. $\bar{X}$ and S

    b. X-mR

    c. u

    d. p

Use the house of quality diagram below to answer questions 75 and 76.

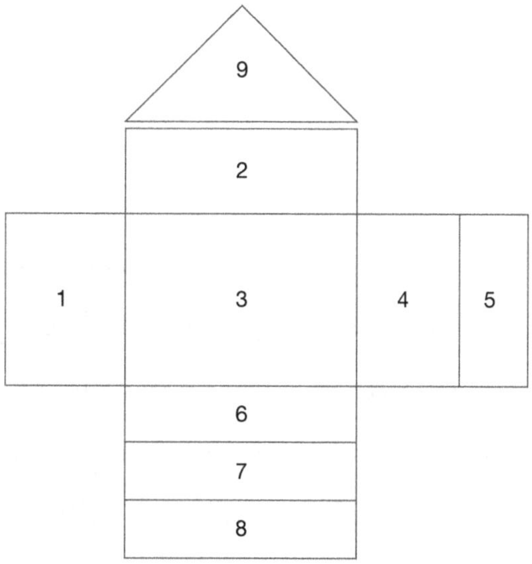

75. The relationship matrix appears in which area of the house of quality?

    a. Area 2

    b. Area 3

    c. Area 6

    d. Area 9

76. The relationship between technical requirements is shown in which area of the house of quality?

    a. Area 1

    b. Area 2

    c. Area 4

    d. Area 9

77. Increased cycle times, reduced throughput, and increased work in process are symptoms of:

    a. the hidden factory.

    b. poor workmanship.

    c. lean implementation.

    d. total productivity maintenance.

78. A 2^{7-3} fractional-factorial experiment has how many fewer runs than a 2^7 full factorial?

    a. 8

    b. 12

    c. 16

    d. 112

79. The Design for Six Sigma procedure DMADOV stands for:

    a. Define, measure, analyze, design, optimize, and verify.

    b. Design, measure, analyze, define, originate, validate.

    c. Define, measure, analyze, design, optimize, and validate.

    d. Design, measure, analyze, define, originate, verify.

80. A disadvantage of systematic sampling is that:

    a. samples may become excessively large.

    b. there is a lack of objectivity.

    c. bias may result if units display a natural cyclic pattern.

    d. identifying units to sample may be difficult.

81. A process output is normally distributed with a mean of 53.70 and standard deviation equal to 8.23. What value for the lower specification limit will result in 5% of the units falling below the specification?

    a. 40.16

    b. 49.00

    c. 67.24

    d. 69.83

82. A common metric in the innovation and learning category of a balanced scorecard is:
    a. cash flow.
    b. cycle time.
    c. employee satisfaction.
    d. defect rates.

83. Trial control limits are calculated based on at least:
    a. 5 subgroups.
    b. 25 subgroups.
    c. 15 subgroups.
    d. 50 subgroups.

84. A company is surveying its 1264 customers. What is the minimum number of survey responses required to achieve a margin of error equal to ± 4%?
    a. 24
    b. 25
    c. 418
    d. 419

85. What is an acceptable measure of dispersion to use for data that is on the nominal measurement scale?
    a. Standard deviation
    b. Range
    c. Mode
    d. None is acceptable.

86. One of the drawbacks of internal benchmarking is that it:
    a. may give competitors access to sensitive data.
    b. is limited by the company's culture, norms, and history.
    c. creates competition among departments.
    d. depends on data that is difficult to collect.

87. Heijunka methods offer a tradeoff between leveling production and:

    a. producing defects.

    b. increasing cycle time.

    c. creating inventory.

    d. reducing availability.

88. Two thousand parts were produced. Sixty-two were sent off for rework and 17 were scrapped. What is the first pass yield?

    a. 0.9605

    b. 0.9615

    c. 0.9685

    d. 0.9915

89. In which role in the RACI model is the individual ultimately held responsible for results?

    a. Responsible

    b. Accountable

    c. Consulted

    d. Informed

90. A gage R&R study was conducted on a tensile tester. The equipment variation is equal to 2.01, and the gage R&R equals 3.20. What is the value of the reproducibility?

    a. 0.84

    b. 1.19

    c. 2.49

    d. 6.20

91. Temperature recorded in degrees Celsius is an example of which scale of measurement?

    a. Nominal

    b. Interval

    c. Ratio

    d. Ordinal

92. A quality inspector chooses every 10th pallet and then samples from the first layer of the chosen pallet. What type of sampling scheme is being used?

    a. Simple random sampling

    b. Block sampling

    c. Systematic sampling

    d. Stratified sampling

93. What type of error occurs when the null hypotheses is not rejected when it is false?

    a. Alpha

    b. Beta

    c. Gamma

    d. No error; it is a correct decision.

94. A Black Belt is studying an X-bar *R* chart that tracks length. What is the status of the X-bar *R* chart shown below?

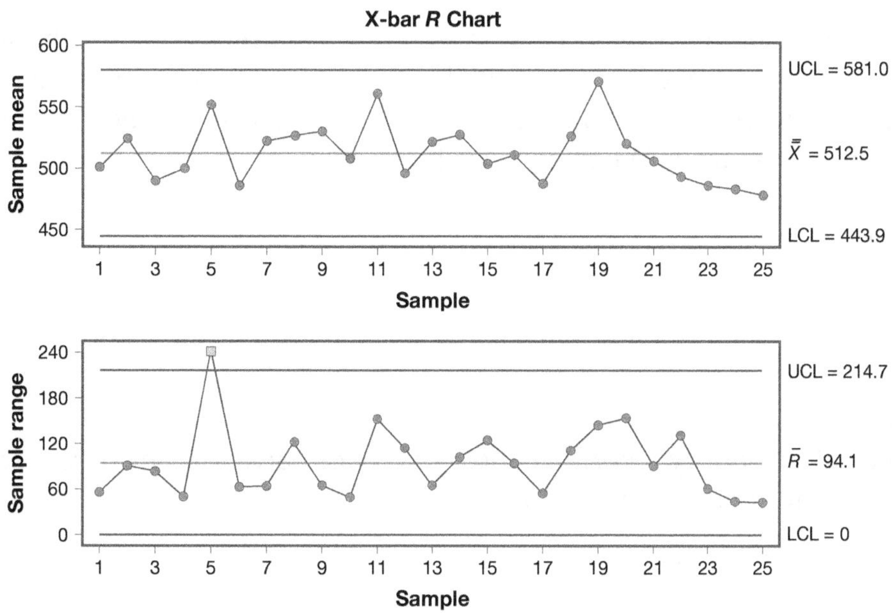

    a. Both the X-bar and *R* charts are out of specification.

    b. The X-bar chart is in control; the *R* chart is out of control at point 5.

    c. Both the X-bar and *R* charts exhibit out-of-control conditions.

    d. The X-bar chart is out of control; the *R* chart is out of specification.

95. A lagging indicator:

    a. shows the number of backorders of a product.

    b. is reported after there is a means to change the outcome.

    c. is forward looking.

    d. can only be operational.

96. The symbol below is used to on a value stream map to denote which of the following items?

    a. Supermarket

    b. Production kanban

    c. Customer

    d. Safety or buffer stock

97. The average customer demand across a shift is 900 units. A shift consists of eight hours with staggered 30-minute lunch breaks during which the line is still running. There are two 15-minute breaks during each shift when the line is shut down for maintenance. What is the takt time in seconds?

    a. 0.50

    b. 0.93

    c. 30.0

    d. 35.8

98. A tool used to identify a performance difference between the current state and a desired future state is known as a:

    a. value stream analysis.

    b. gap analysis.

    c. SMART goal.

    d. key performance indicator.

99. A team member is cooperative and assertive. They are likely to be:

    a. competing.

    b. accommodating.

c. compromising.

d. collaborating.

100. The phrase "vital few and useful many" is applicable to the:

a. cause-and-effect diagram.

b. check sheet.

c. Pareto diagram.

d. process flow diagram.

101. Which of the following items is an example of soft savings?

a. Savings that are false

b. Increased customer satisfaction

c. Increased throughput

d. Reduced supplier costs

102. A Black Belt performs a hypothesis test, and the analysis returns a p-value equal to 0.075. What is the interpretation of the p-value?

a. The null hypothesis is true with a probability equal to 0.925.

b. The alternative hypothesis is true with a probability equal to 0.075.

c. Assuming H_0 is true, the probability of the current result or something more extreme is 0.075.

d. Assuming H_1 is true, the probability of the current result or something more extreme is equal to 0.925.

103. Which of the following statements concerning the p-chart is true?

a. It applies to quality characteristics that are continuous measures.

b. It gives a poor overall quality picture.

c. It can be used to track defects.

d. It can be used for varying sample sizes.

104. Which leadership approach is best used for team members with medium competence but low commitment?

a. Coaching

b. Delegating

c. Directing

d. Supportive

Use the figure below to answer questions 105 and 106.

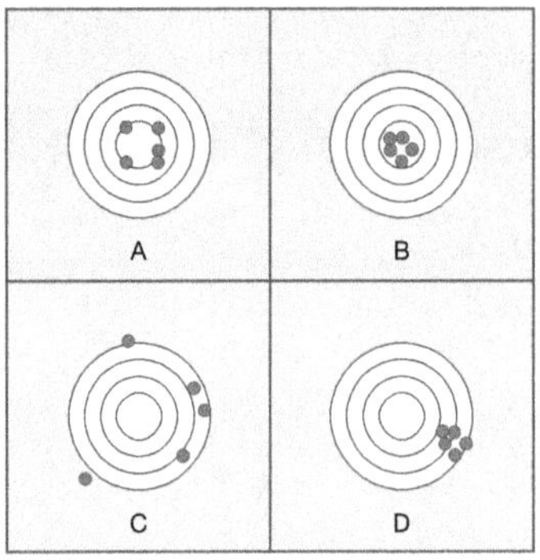

105. Which quadrant of the figure shows low precision and low accuracy?

   a. Quadrant A

   b. Quadrant B

   c. Quadrant C

   d. Quadrant D

106. Which quadrant in the figure shows high precision and low accuracy?

   a. Quadrant A

   b. Quadrant B

   c. Quadrant C

   d. Quadrant D

107. OEE is a metric that combines which three measures?

   a. Availability, productivity, and quality

   b. Availability, cycle time, and gage R&R

   c. Takt time, inventory, and defects

   d. Work in process, cycle time, and quality

108. Given the two-way ANOVA table below, what is the *F*-statistic for the AB interaction?

Source of Variation	**Sum of Squares**	**Degrees of Freedom**	**Mean Square**	***F*-statistic**
Factor A	344	3		
Factor B	234	4		
AB Interaction	120			
Error				
Total	**775**	**24**		

a. 0.65

b. 1.54

c. 2.23

d. 2.86

109. In a supplier risk analysis, items that are classified as low supply-risk, but high profit-impact are considered:

a. noncritical items.

b. bottleneck items.

c. strategic items.

d. leverage items.

110. What sets Six Sigma projects apart from other types of projects?

a. Six Sigma projects cost more to implement.

b. In a Six Sigma project, the solution is not known in advance.

c. Six Sigma projects use teams.

d. Six Sigma projects use Gantt charts to track progress.

111. SMED can directly help increase which component of OEE?

a. Availability

b. Performance

c. Quality

d. Kaizen

112. Calculate the upper control limit for the *S* chart given the following information: X-bar = 904.0, S-bar = 32.5, and $n = 20$.

a. 909.9

b. 920.9

c. 926.1

d. 952.4

113. Before an improvement project, the percentage defective for a milling process was equal to 4.9%. After the changes were made, 100 parts were run, and the percentage defective was equal to 4.0%. Use the normal approximation method to test if the decrease in percentage defective is statistically significant at the 95% level.

    a. $z_0 = -0.42$, $z_{crit} = -1.645$; do not reject H_0.

    b. $z_0 = -0.46$, $z_{crit} = -1.645$; do not reject H_0.

    c. $z_0 = -0.42$, $z_{crit} = -1.96$; do not reject H_0.

    d. $z_0 = -0.46$, $z_{crit} = -1.96$; do not reject H_0.

114. Once an organization identifies a risk, it can choose one of which four actions?

    a. Target, terminate, treat, or transfer

    b. Tolerate, terminate, treat, or transfer

    c. Temper, treat, transfer, or tighten

    d. Tolerate, tailor, transfer, or treat

115. Which of the following categories in a SWOT analysis are considered internal to the organization?

    a. Opportunities and strengths

    b. Weaknesses and threats

    c. Opportunities and threats

    d. Strengths and weaknesses

116. A job shop makes three metal parts. A Black Belt tests whether last month's product mix differed from last year's production percentages. What is the value of the test statistic and critical value assuming alpha = 0.05?

	Part A	Part B	Part C
Last year's percentage	25%	50%	25%
Last month's observed units	340	600	360
Last month's expected units	325.0	650.0	325.0

    a. Chi-sq = 8.308, Chi-square critical = 11.070

    b. Chi-sq = 8.231, Chi-square critical = 7.378

    c. Chi-sq = 8.308, Chi-square critical = 5.991

    d. Chi-sq = 8.231, Chi-square critical = 12.833

117. Work breakdown structures serve as inputs into tools such as the:

   a. affinity diagram.

   b. Gantt chart.

   c. FMEA.

   d. control plan.

118. In a Resolution V experiment:

   a. some main effects are confounded with two-factor interactions.

   b. some main effects are confounded with three-factor interactions.

   c. some two-factor interactions are confounded with three-factor interactions.

   d. some two-factors interactions are confounded with the mean.

119. Which leadership style is best used for a team member that has low competence but high commitment?

   a. Delegating

   b. Coaching

   c. Directing

   d. Supporting

120. Which of the Big Losses are associated with equipment performance?

   a. Breakdowns and setup and changeover time

   b. Idling and minor stoppages

   c. Startup losses

   d. Defects, rework, and startup losses

121. To respond to a cybersecurity threat, an organization may use which strategy?

   a. Continuously hunt for network intrusions.

   b. Update and upgrade all software immediately.

   c. Leverage modern hardware security features.

   d. Transition to multifactor authentication.

122. A new electric vehicle is evaluated for its maximum range under a single charge. Five vehicles were tested on a track, with the mileage results shown below. Assuming mileage is normally distributed, test whether the average range is at least 250 miles using an alpha = 0.05.

{265, 252, 255, 260, 251}

a. $t_0 = 1.13$; do not reject H_0.

b. $z_0 = 1.13$; do not reject H_0.

c. $t_0 = 2.52$; reject H_0.

d. $z_0 = 2.52$; reject H_0.

123. What might be the effect of using 2-sigma control limits on an X-bar chart?

a. The alpha risk will increase.

b. The beta risk will increase.

c. The capability index will decrease.

d. Special cause variation will decrease.

124. A Black Belt conducts a hypothesis test using a Student's t statistic and $\alpha = 0.05$. The p-value of the test is equal to 0.325. Which of the following statements is true?

a. The null hypothesis is true with probability 0.675.

b. The probability of the alternative hypothesis being true is 0.325.

c. There is insufficient evidence to reject the null hypothesis.

d. There is sufficient evidence to accept the null hypothesis.

125. After making adjustments to a packaging line, a Six Sigma professional wants to determine with 95% confidence whether the mean fill weight differs by 1 g. The process historical standard deviation is $\sigma = 3.5$ g. What is the minimum required sample size needed to make this determination?

a. 21

b. 25

c. 34

d. 48

126. In an autocratic (A2) decision-making approach:

    a. the team is consulted to obtain specific information and then the leader makes the final decision.

    b. the team is informed of the situation and each member's opinion is asked for individually. The group is not brought together. The leader then makes the final decision.

    c. The leader works with the team to reach a group consensus.

    d. The team is brought together to discuss the issue and make suggestions and the leader then makes the final decision.

127. The 95% confidence interval [12.5, 18.5] was calculated on the mean of a population with a known standard deviation, σ. The sample size was $n = 20$. What is the value of σ?

    a. 3.61

    b. 6.67

    c. 6.85

    d. 30.61

128. A u-chart tracks:

    a. the number of defective units.

    b. the number of defects.

    c. the number of standard deviations.

    d. the defect rate per unit.

129. A process step in a PFMEA has an RPN = 96, severity of 8, and a probability of occurrence of 6. What is the detection score?

    a. 2

    b. 12

    c. 16

    d. 48

130. Schedule boards, signage, and color-coded bins are examples of:

    a. visual management.

    b. kanban.

    c. theory of constraints.

    d. 5S.

131. If a project stakeholder is deemed to have low influence and high importance, what behavior approach should the project leader adopt for this stakeholder?

    a. Collaborate

    b. Engage to prevent sabotage

    c. Protect, defend, and give voice

    d. Monitor

132. Which of the following subgroup sizes for an $\overline{X}$ and R chart would not be used?

    a. 1

    b. 2

    c. 5

    d. 7

133. A Black Belt is testing whether the standard deviations of two normal populations are different with $\alpha = 0.05$. They take samples of sizes $n_1 = 30$ and $n_2 = 25$. What is the critical value for the corresponding hypothesis test?

    a. $F(0.05, 30, 25)$

    b. $\sqrt{}\, F(0.025, 29, 24)$

    c. $F(0.025, 29, 24)$

    d. $\chi^2\ (0.025, 54)$

134. Robust design methods aim to:

    a. set noise factors to minimize the effects of control factors.

    b. set control factors to minimize the effects of noise factors.

    c. set noise factors to optimal levels.

    d. identify noise factors and eliminate them.

135. A production supervisor runs a punch press line for a few extra hours to create back-up stock "just in case." What types of waste have been created?

    a. Defects and skills

    b. Overproduction and inventory

    c. Waiting and transportation

    d. Defects and motion

136. On an eight-hour shift, a process runs a total of 438 minutes due to a scheduled 30-minute shutdown for lunch and 12 minutes of downtime due to a broken clamp. What is the availability of the process?

 a. 0.9125

 b. 0.9315

 c. 0.9375

 d. 0.9733

137. A quality engineer will test a sample of lightbulbs. What is the required minimum sample size to assure a 3% margin of error for a confidence interval on the percentage of defective bulbs?

 a. 266

 b. 267

 c. 1067

 d. 1068

138. An evaluation method that uses a three-dimensional miniature version of a solution is known as a:

 a. physical model.

 b. pilot run.

 c. prototype.

 d. simulation.

139. In a two-level full-factorial design, the effect of Factor A is equal to 5.72. What is the corresponding coefficient for factor A in the coded regression model?

 a. –5.72

 b. 2.86

 c. 5.72

 d. 11.44

140. In a balanced scorecard, which perspective includes the capabilities and skills of an organization?

 a. Financial perspective

 b. Learning and growth perspective

 c. Internal business processes perspective

 d. Customer perspective

141. What experimental design technique would be used when an experiment will have to be conducted over the course of two days?

   a. Confounding

   b. Replication

   c. Randomization

   d. Blocking

142. A university institutes two-factor authentication for logging into its systems. This is an example of:

   a. mistake-proofing.

   b. a bottleneck.

   c. a pull system.

   d. heijunka.

143. A typical error that leaders can make when managing change is to:

   a. provide too much explanation about the change.

   b. assign the change-management function to people without the resources to execute it.

   c. communicate too much information.

   d. move too slowly to implement changes.

144. What is the most common loss function used for a minimum tolerance?

   a. Variance is best.

   b. Smaller is better.

   c. Nominal is best.

   d. Larger is better.

145. A kaizen blitz is best used in which situation?

   a. The risk of failure is high.

   b. Projects are open-ended, with no time constraint.

   c. Projects are not clearly defined.

   d. Improvement opportunities are readily identifiable.

146. Guidelines used in Design for Maintainability include:

    a. designing in self-diagnostics.

    b. using self-locating fixtures.

    c. minimizing life cycle cost.

    d. using exotic materials for manufacturing.

147. List the steps of system improvement using the theory of constraints in order, from first to last.

    1. Exploit
    2. Elevate
    3. Identify
    4. Repeat
    5. Subordinate

    a. 3, 1, 5, 2, 4

    b. 3, 2, 5, 1, 4

    c. 3, 1, 2, 5, 4

    d. 3, 5, 2, 1, 4

148. The upper control limit on a *p*-chart varies over time. This is due to changes in the:

    a. percentage defective.

    b. defect rate.

    c. sample size.

    d. range.

149. A customer survey asks how likely a customer would be to recommend the product to a friend or colleague. The survey uses a Likert scale with 1 = *Not likely at all* to 10 = *Extremely likely*. Customers responding with a 7 or 8 are:

    a. enthusiastic and willing to refer the product to others.

    b. satisfied but unenthusiastic about the product.

    c. confused about the functionality of the product.

    d. unhappy and may damage the brand by posting negative reviews.

150. Which of the following terms is considered the fourth stage in a Six Sigma maturity model?

   a. Launch

   b. Culture transformation

   c. Scale and replication

   d. Institutionalization

SOLUTIONS

1. d; Some hallmarks of Six Sigma include the use of teams to work on well-defined projects and an emphasis on a five-step problem-solving process. [I.A.1]

2. b; A system that is designed to implement a process or set of processes is known as a business system. [I.A.3]

3. d; An effective control plan addresses monitoring, detecting, diagnosing, and reaction plans. [VIII.D.2]

4. a; Learning and performance differ in that learning creates potential, and performance is the realization of that potential. [III.D.2]

5. a; To calculate the tolerance limit that covers at least 99% of the distribution with 95% confidence, we can use Appendix J.2 in *The ASQ Certified Six Sigma Black Belt Handbook*. For $n = 50$, the two-sided tolerance factor is $K = 3.117$. Then we calculate the tolerance limit to be

$$[\bar{X} - Ks, \bar{X} + Ks] = [1.5 - 3.117 \times 0.20, 1.5 + 3.117 \times 0.20]$$
$$= [0.877, 2.123].$$

   [VI.B.4]

6. b; Using the conventional method, the total tolerance for the assembly is the sum of the tolerance of the parts:

$$Tolerance_{Total} = 0.05 + 0.10 + 0.50 = 0.65.$$

   [IX.C]

7. b; With the consultative decision-making approach, the team is informed of the situation and each member's opinion is asked for individually. The group is not brought together. The leader then makes the final decision. [III.C.3]

8. c; The first step in developing a training plan should be to assess the need for training. [III.D.1]

9. d; Refilling the lubricant is an example of condition-based maintenance, in which maintenance tasks are performed based on operating conditions. [VIII.B.1]

10. c; Virtual teams are least affected by geography. [III.A.1]

11. b; A balanced designed experiment is one in which the number of runs for each factor level is equal. [VII.A.2]

12. c; Work in progress (WIP) is the material that has entered the process but has not reached the output or finished stage. Work in queue is a component of WIP. [V.A.1]

13. b; Given the precedence table below, we can draw the activity network diagram and find the critical path.

Activity	Predecessor	Time
A	—	10
B	—	5
C	A, B	10
D	—	5
E	D	5
F	E	5
G	C, F	5

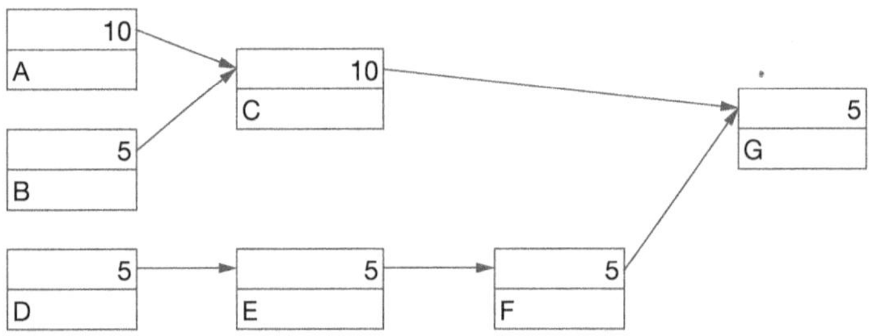

There are three paths: ACG = 25, BCG = 20 and DEFG = 20. The path with the longest duration is the critical path, which is ACG. [IV.D.5]

14. c; Activity D is not on the critical path. The slack in path DEFG is equal to 25 – 20 = 5 days. Therefore, the late start for Activity D is 5. [IV.D.5]

15. b; White Belts have a basic understanding of Six Sigma principles. [I.B.1]

16. c; By simply concentrating on speed, the team may inadvertently create more process errors. Tracking both processing time and error counts will help the team avoid suboptimizing the process. [IV.B.4]

17. a; An employee who might experience negative effects from a Six Sigma project would be considered a stakeholder. Stakeholders are those individuals or entities that will be affected by a project or can influence the outcome of the project. [II.A]

18. d; Design failure mode and effects analysis (DFMEA) is a tool used in the design process to minimize or prevent failures and improve safety and quality. [VI.C.2]

19. a; A process decision program chart (PDPC) is a type of tree diagram. [IV.D.6]

20. b; An *np*-chart is used to track the number of defectives (failed units) in a sample when the sample size is constant over time. [VIII.A.4]

21. b; The statistical standard for quality associated with Six Sigma is 3.4 defects per million opportunities. This assumes that a 6-sigma, normally distributed process centered at target has shifted its mean by 1.5 standard deviations. [I.A.1]

22. c; Weekly average call length is an example of a tactical performance indicator. A tactical performance indicator is a consolidated metric used by first-level managers to track progress and make decisions. [II.C.1]

23. b; Cycle time is the time required to complete one unit from the beginning of the process to the end of the process. [V.A.1]

24. c; A decrease in project funding can result in a decrease in time and a decrease in scope. There is an interplay among cost, quality, scope, and time called the balanced quadrant. [IV.B.5]

25. d; When studying the effects of various percentages of ingredients, use a mixture design. These designs incorporate the dependency of the ingredient percentages in that they must sum to 100%. [VII.A.1]

26. c; Takt time is the rate at which units must be completed to meet customer demand. [V.A.1]

27. a; Time zones can be a constraint on the effectiveness of a virtual team. [III.A.1]

28. b; SIPOC stands for suppliers—inputs—process steps—outputs—customers. C = customers. [IV.A.3]

29. d; The net promoter score is calculated based on the number of customers who respond that they would recommend the brand or product to a friend or colleague. [II.C.1]

30. a; SMED, single-minute exchange of dies, is also known as quick changeover. The technique emphasizes the use of items such as guide rails, quick-change fasteners, staging carts, and interchangeable parts. [VII.B.2]

31. c; In the performing phase of team development, members understand one another and recognize each other's strengths and weaknesses. [III.B.2]

32. a; Involving stakeholders in project charter reviews helps ensure cooperations. [IV.B.6]

33. b; Opportunity identification and project evaluation are categories that are typically evaluated in a Six Sigma maturity model. [I.A.1]

34. b; The nodes in an interrelationship digraph with the most incoming arrows are considered outcomes. Nodes with the most outgoing arrows are considered drivers. [IV.D.7]

35. a; A stakeholder who has been designated as high potential for cooperation and high potential to impact a project would be considered a mixed blessing. [II.A]

36. a; In a tollgate review for the analyze phase, a champion may ask "What are the critical Xs?" The Analyze phase allows the team to make the connection between outputs and important inputs. [IV.C.2]

37. c; Color-coding, floor markings, and indictor lights are examples of visual management. [VIII.B.2]

38. c; If a project champion concludes that the project is no longer consistent with the goals of the organization, the project should be canceled. [IV.C.2]

39. b; If performance data only on products that have been in service for at least 1000 hours is collected, the resulting sample is not likely to be representative of the population due to survivorship bias. [V.B.3]

40. a; When addressing disruptive behavior in a meeting, the team leader should first make eye contact with the individual. Often, this is enough to stop the behavior. If not, the leader can choose to escalate their response. [III.C.1]

41. d; Independent events have no influence over each other. [V.E.1]

42. c; An asset security continuity plan may include the loss of intellectual property. [I.A.4]

43. c; A work breakdown structure is used to show how smaller activities are linked to larger activities in a project. This information can then be used as an input to a Gantt chart. [IV.C.3]

44. a; A spaghetti diagram that displays several intersecting lines indicates that the department layout should be redesigned. [V.A.2]

45. a; Once the part-to-part variation is decreased, an improvement team should reevaluate the measurement system. If the measurement system variation has not changed, its percentage of total variation may increase to unacceptable levels. [VIII.C.1]

46. b; The phase of team development in which team members express their own opinions and ideas, often disagreeing with others, is referred to as storming. [III.B.2]

47. a; The rolled throughput yield for the four steps in series is equal to the product of the FTY of each step. [V.F.6]

$$RTY = 0.98 \times 0.945 \times 0.90 \times 0.97 = 0.808$$

48. c; A good problem statement should contain where the problem is occurring, how long it has existed, the measurable item, the performance gap, and the business impact. The statement is missing the performance gap and the business impact. [IV.B.2]

49. d; Data on the interval scale has meaningful differences but no true zero point. Examples are temperatures in Celsius and Fahrenheit, and years. [V.B.2]

50. c; Multivoting is a technique used to select items from a brainstormed list using point assignments. [III.C.3]

51. b; Lean is an improvement philosophy that is concerned with reducing waste. [I.A.2]

52. b; The minimum number of runs needed for a full factorial experiment with three factors at two levels each is $2^3 = 8$. [VII.A.6]

53. a; When benchmarking, the final step a company should take is to revise internal performance measurements and goals based on what it learns from the exercise. [II.B]

54. c; If a two-factor interaction graph from a full-factorial experiment displays two lines that are parallel, it indicates that there is no significant interaction between the factors. [VII.A.6]

55. b; In statistical process control, the aim is to minimize within-subgroup variation to make between-subgroup variation more prominent. [VIII.A.3]

56. b; A T-shaped matrix diagram can be used to show pairwise relationships between three variables. [IV.D.3]

57. c; A team leader may give out action assignments to each member to reduce their reluctance to participate. [III.C.1]

58. b; Work instructions are an important input in standardized work. [V.A.2]

59. d; Fault tree analysis is a root cause analysis method that uses AND and OR gates. The tool helps the team consider underlying causes for a failure model and study relationships between various failures. [VI.D.2]

60. d; Key behavior indicators (KBIs) can be used to track cultural change in an organization. [II.C.1]

61. c; Organization should concentrate on process efficiency for those items considered noncritical in the supplier risk matrix. [VI.C.1]

62. d; To find the PPM, we use

$$PPM = \frac{Defectives}{Total} \times 10^6$$
$$= \frac{244}{32{,}678} = 0.007466797 \times 10^6 = 7466.8.$$

[V.F.6]

63. b; Because they allow for varying sample sizes, the p- and u-charts can have control limits that vary in width. The larger the sample size, the narrower the control limits. [VIII.A.4]

64. d; During a tollgate meeting for the measure phase of a project, a champion may ask, "What did the FMEA show?" [IV.C.2]

65. c; The probability of finding four or more defects is expressed as $\Pr(X \geq 4)$. The problem would then be solved using the Poisson distribution. [V.E.2]

66. b; Takt time can be thought of as the rate of customer pull. [VII.B.2]

67. a; Six Sigma projects should be consistent with an organization's long-term strategic objectives and customer requirements. [I.A.2]

68. a; In this situation, the R&D department would likely use Design for Cost. [IX.B]

69. d; To find the first pass yield (FPY) from defects per unit (DPU), we use the Poisson distribution with mean = dpu to find $\Pr(X = 0)$:

$$\Pr(X = 0) = \frac{e^{-\mu}\mu^{x}}{x!} = \frac{e^{-dpu}dpu^{0}}{0!} = e^{-0.40} = 0.67.$$

[V.F.6]

70. c; When a high-performing team undergoes changes to its participating members, the team often regresses at least one stage in the team development process. In this case, the team would likely regress to the norming phase. [II.B.2]

71. a; Adding center points to a factorial design has many advantages. One of these is that it helps the analyst detect changes in the experimental environment over time. In addition, the incorporation of center points allows for testing for curvature in the model, adds to the degrees of freedom of the analysis, and gives an estimate of pure error. [VII.A.4]

72. d; Delighters in Kano analysis are not expected by the customer, so their absence will not adversely affect customer satisfaction. Adding a delighter will increase customer satisfaction. [IV.A.3]

73. c; Listing out the units in a population and then randomly choosing n units is an example of simple random sampling. [V.B.3]

74. d; A circuit board manufacturing line requires continuous process monitoring using a count of defective items. Since the sample size will vary, the *p*-chart is applicable here. [VIII.A.4]

75. b; The relationship matrix appears in Area 3 of the house of quality. [IV.A.3]

76. d; The pairwise relationships between technical requirements appear in Area 9, the roof of the house of quality. [IV.A.3]

77. a; Symptoms of the hidden factory include increased cycle times, reduced throughput, and increased work in process. [V.A.1]

78. d; A 2^7 full-factorial experiment has 128 runs. The 2^{7-3} fractional factorial has $2^4 = 16$ runs. The difference is 128 – 16 = 112 fewer runs. [VII.A.5]

79. a; The Design for Six Sigma procedure DMADOV stands for define, measure, analyze, design, optimize, and verify. [IX.A]

80. c; A disadvantage of systematic sampling is that bias may result if units display a natural cyclic pattern. The systematic samples may capture the same position on the cycle each time. [V.B.3]

81. a; To set the lower specification limit so that it results in 5% of the units falling below the spec, we find the critical value $z_{0.05} = -1.645$ from the standard normal table. Then we solve for X using

$$Z = \frac{(X - \mu)}{\sigma}$$

$$-1.645 = \frac{(X - 53.70)}{8.23}$$

$$X = 53.70 - 1.645 \times 8.23$$

$$X = 40.16.$$

[V.E.2]

82. c; Employee satisfaction is a common metric in the innovation and learning category of a balanced scorecard. [II.C.1]

83. b; Trial control limits for an SPC chart are calculated based on at least 25 subgroups. [VIII.A.4]

84. d; For a known population size, we calculate the minimum sample size needed to achieve an $X\%$ margin of error using

$$\lceil n = \frac{N}{1 + NE^2} = \frac{1264}{1 + (1264)0.04^2} = 418.21 \uparrow 419.$$

[V.B.4]

85. d; There is no acceptable measure of dispersion for data that is on nominal measurement scale. Nominal data cannot be ordered in any meaningful mathematical way. The measure of central tendency that can be used for nominal data is the mode. [V.B.2]

86. b; One of the drawbacks of internal benchmarking is that it is limited by the company's culture, norms, and history. [II.B]

87. c; Heijunka methods offer a tradeoff between leveling production and creating inventory. [VII.B.2]

88. a; First pass yield is calculated using

$$FPY = \frac{Total - rework - scrap}{Total} = \frac{2000 - 62 - 17}{2000} = 0.9605.$$

[V.F.7]

89. b; The "accountable" role in a RACI diagram is given to the individual who will ultimately be held responsible for results. This role is often given to upper management. [IV.C.4]

90. c; The gage R&R calculation involves the equipment variation (repeatability) as well as the appraiser variation (reproducibility). We use the formula for gage R&R and solve for the appraiser variation, or reproducibility:

$$\sigma_{Gage\ R\&R} = \sqrt{EV^2 + AV^2}$$
$$3.20 = \sqrt{2.01^2 + AV^2}$$
$$AV^2 = \left(3.20^2 - 2.01^2\right) = 6.20$$
$$AV = \sqrt{6.20} = 2.49.$$

[V.C.1]

91. b; Temperature measured in either Celsius or Fahrenheit is an example of an interval measurement scale. The values can be ordered, and the difference between values is meaningful, but there is no true zero. [V.B.2]

92. c; Choosing every 10th pallet and then sampling from the first layer of the pallet is an example of systematic sampling. [V.B.3]

93. b; If the null hypothesis is not rejected when it is false, this is an error of inaction, also known as a beta or Type II error. [VI.B.1]

94. c; The X-bar chart is out of control due to six points in row trending down. The *R* chart is also out of control at point 5. The X-bar *R* chart tracks whether the process in is statistical control, not whether the units are within specifications. [VIII.A.5]

95. b; A lagging indicator is reported after there is a way to change the outcome. It is preferrable to use leading indicators to be able to detect problems earlier instead of after the fact. [II.C.1]

96. b; The symbol below is used to denote a production kanban in a value stream map. [V.A.2]

97. c; The net available time during the shift is 8 × 60 = 480 minutes, less the two 15-minute breaks for maintenance. The line is not shut down for lunch. Finding takt time in seconds, then

$$\begin{aligned} \textit{Takt time} &= \frac{\textit{Net available time}}{\textit{Customer demand}} \\ &= \frac{(8\times 60-15-15)\times 60}{900} \\ &= 30.0\,\text{sec/unit}. \end{aligned}$$

[VII.B.2]

98. b; Gap analysis is a tool used to identify a performance difference between the current and a desired future state using metrics implied by the states. [VI.D.1]

99. d; A team member who is cooperative and assertive is likely to be a collaborating member of a team. [III.C.1]

100. c; The Pareto diagram is based on the concept of the vital few and the useful many. [VI.D.2]

101. b; Increased customer satisfaction is a project outcome that is often difficult to assign a hard dollar amount to; it would be considered "soft savings." [I.A.4]

102. c; The *p*-value is calculated under the assumption that the null hypothesis is true. It is the probability of getting the current test statistics or one that is more extreme assuming that H_0 is true. A *p*-value less than alpha will prompt us to reject H_0 and conclude that H_1 is correct. [VI.B.1]

103. d; The *p*-chart can be used for tracking defectives when sample size varies. [VIII.A.4]

104. a; For team members with moderate competence but low commitment, the team leader should use the coaching approach. [III.B.4]

105. c; Quadrant C displays low precision and low accuracy. [V.C.1]

106. d; Quadrant D displays high precision and low accuracy. [V.C.1]

107. a; Overall equipment effectiveness, or OEE, combines availability, productivity, and quality. OEE = A × P × Q. [VII.B.4]

108. a; The completed ANOVA table is shown below:

Source of Variation	Sum of Squares	Degrees of Freedom	Mean Square	F-statistic
Factor A	344	3	114.67	7.45
Factor B	234	4	58.5	3.80
Interaction	120	**12**	**10**	**0.65**
Error	**77**	5	**15.4**	
Total	775	24		

$$SS_E = SS_{Total} - SS_A - SS_B - SS_{AB}$$
$$= 775 - 344 - 234 - 120 = 77$$

$$DF_{AB} = DF_A \times DF_B = 3 \times 4 = 12$$
$$DF_{Error} = DF_{Total} - DF_A - DF_B - DF_{AB}$$
$$= 24 - 3 - 4 - 12 = 5$$

$$MSE_{AB} = \frac{SS_{AB}}{DF_{AB}} = \frac{120}{12} = 10$$

$$MSE = \frac{SS_{Error}}{DF_{Error}} = \frac{77}{5} = 15.4$$

$$F_{AB} = \frac{MS_{AB}}{MS_{Error}} = \frac{10}{15.4} = 0.65.$$

[VI.B.6]

109. d; In a supplier risk analysis, items that are classified as low supply-risk but high profit-impact are considered leverage items. [VI.C.1]

110. b; In a Six Sigma project, the solution is not known in advance. This can add complexity to the project management effort. [IV.B.5]

111. a; SMED, or quick changeover, reduces set-up time, thus increasing equipment availability. [VII.B.4]

112. c; The formula for the upper control limit of the X-bar chart in an X-bar/S chart requires a lookup table value A_3 based on the subgroup size n. For $n = 20$, $A_3 = 0.680$. We then use

$$UCL_{\overline{X}} = \overline{X} + A_3\overline{S} = 904.0 + 0.680 \times 32.5 = 926.1.$$

[VIII.A.4]

Part B Solutions

113. a; Here we test the new percentage defective versus the previous percentage of 4.9%. If the improvement was effective, the new percentage will be significantly less than the original p_0. The hypotheses for this test are

$$H_0: p_{new} = 0.49 \text{ vs. } H_1: p_{new} < 0.49.$$

We note that the assumptions for this test—namely that $np_0 \geq 10$ and $n(1 - p_0) \geq 10$—are both satisfied. Using the normal approximation method, we calculate the test statistic:

$$z_0 = \frac{(\hat{p} - p_0)}{\sqrt{\frac{p_0(1-p_0)}{n}}} = \frac{(0.04 - 0.049)}{\sqrt{\frac{0.049(1-0.049)}{100}}} = -0.42.$$

We then compare this to the critical value $Z_{0.05} = -1.645$. Since the calculated z_0 is not less than Z_{crit}, we fail to reject H_0. We conclude there is insufficient evidence to claim the percent defective rate has decreased. [VI.B.5]

114. b; Once an organization identifies a risk, it can choose to tolerate, terminate, treat, or transfer the risk. [VI.C.1]

115. d; The acronym SWOT stands for strengths, weaknesses, opportunities, and threats. The categories that are considered internal to the organization are strengths and weaknesses. [I.A.4]

116. c; This problem requires a chi-square goodness-of-fit test. Here,

H_0: Last month's mix follows the historical percentages.
H_1: Last month's mix differs from the historical percentage.

We calculate the test statistic using

$$\begin{aligned}\chi_0^2 &= \sum \frac{(Observed - Expected)^2}{Expected} \\ &= \frac{(340-325)^2}{325} + \frac{(600-650)^2}{650} + \frac{(360-325)^2}{325} \\ &= 8.308.\end{aligned}$$

We find the critical value of 5.991 from the Chi-square table using $(k - 1) = 2$ degrees of freedom and $\alpha = 0.05$. Since the value of the calculated test statistic is greater than the critical value, we reject H_0 and claim that the product mix has changed. [VI.B.7]

117. b; Work breakdown structure serves as inputs to a Gantt chart. [IV.C.3]

118. c; In a Resolution V design, some two-factor interactions are confounded with three-factor interactions. [VII.A.5]

119. c; For team members with low competence but high commitment, the directing leadership approach is best. [III.B.4]

120. b; The losses of idling and minor stoppages are associated with equipment performance. [VII.B.4]

121. a; An organization may continuously hunt for network intrusions to respond to a cybersecurity threat. [VI.C.1]

122. c; We conduct a one-sided hypothesis test on the mean and use a Student's t test statistic. The hypotheses are shown below:

$$H_0: \mu \leq 250$$

$$H_1: \mu > 250.$$

Since we do not know the population standard deviation, we will use a Student's t test. To calculate the Student's t statistic, we first note that $n = 5$ and find the sample mean and standard deviation:

$$\bar{X} = \frac{\sum x_i}{n} = \frac{(265 + 252 + 255 + 260 + 251)}{5} = 256.60$$

$$s = \sqrt{\frac{\sum x_i^2 - n\bar{x}^2}{n - 1}} = \sqrt{\frac{329355 - 5 \times 256.6^2}{4}} = 5.86.$$

Then

$$t_0 = \frac{(\bar{X} - \mu_0)}{\frac{s}{\sqrt{n}}} = \frac{(256.6 - 250)}{\frac{5.86}{\sqrt{5}}} = 2.52.$$

We compare this test statistic to a critical value from the Student's t table with alpha = 0.05 and $(n - 1) = 4$ degrees of freedom. The value $t(0.05, 4) = 2.132$. Because $t_0 > t_{crit}$, we reject H_0 and conclude that the average mileage is greater than 250. [VI.B.5]

123. a; If 2-sigma control limits are used on an X-bar chart, the false alarm rate, or alpha error, will increase from 0.0027 to 0.0455. The average run length (ARL) before a false alarm for 3-sigma limits is $1/\alpha = 1/0.0027 = 370$. For 2-sigma limits, it is $1/\alpha = 1/0.0455 = 22$. [VIII.A.4]

124. c; For a $p = 0.325$, the correct wording of the conclusion is that there is insufficient evidence to reject the null hypothesis. We neither prove nor disprove a null hypothesis. The p-value for a hypothesis test is defined as:

$$\Pr(\text{The test statistic value or one more extreme} \mid H_0 \text{ is true}).$$

A p-value greater than α does not imply that the null is true; we say we have insufficient evidence to reject the null.
[VI.B.1]

125. d; The specified margin of error for the population mean is equal to 1 g. We find the minimum sample size for a 95% two-sided confidence interval using

$$n \geq \left(\frac{z_{\alpha/2}\sigma}{E}\right)^2$$
$$\geq \left(\frac{1.96 \times 3.5}{1}\right)^2$$
$$\geq 47.06 = 48.$$

[VI.B.3]

126. a; In an autocratic (A2) decision-making approach, the team is consulted to obtain specific information and then the leader makes the final decision. [III.C.3]

127. c; The confidence interval [12.5, 18.5] was calculated using

$$\bar{X} \pm E$$
$$\bar{X} \pm z_{\alpha/2}\frac{\sigma}{\sqrt{n}}.$$

Using the end points of the interval, we solve for the margin of error, E:

$$E = \left(\frac{18.5 - 12.5}{2}\right) = 3.0.$$

Now, we note that $z_{\alpha/2} = 1.96$ and solve for sigma:

$$3.0 = 1.96\frac{\sigma}{\sqrt{n}} = 1.96\frac{\sigma}{\sqrt{20}}$$

$$\sigma = \frac{3.0\sqrt{20}}{1.96}$$
$$= 6.85.$$

[VI.B.4]

Part B Solutions

128. d; A u-chart tracks the defect rate per unit. [VIII.A.4]

129. a; The risk priority number (RPN) is calculated as *Severity* × *Occurrence* × *Detection*:

$$Detection = RPN \div (Severity \times Occurrence) = 96 \div (8 \times 6) = 2.\ [VI.C.2]$$

130. a; Schedule boards, signage, and color-coded bins are examples of visual management. [VIII.B.2]

131. c; If a project stakeholder is deemed to have low influence and high importance, the project leader should adopt the "protect and defend, and give voice" approach. [II.A]

132. a; The X-bar and R chart uses between two to nine samples per subgroup. If there is a single sample per time period, Black Belts can use an X-mr chart or an EWMA chart. [VIII.A.4]

133. c; A hypothesis test on the ratio of two standard deviations from normal distributions uses the F distribution. The hypothesis test described is two-sided, so we will divide the alpha value by 2 when looking up the critical value in the F table. We use the degrees of freedom $(n_1 - 1) = 29$ and $(n_2 - 1) = 24$. The look up value is then $F(0.025, 29, 24)$. [VI.B.5]

134. b; Robust design aims to set control factors to minimize the effects of noise factors. [IX.C]

135. b; Running the line a few extra hours to build up inventory would create both overproduction and inventory waste. [VI.D.3]

136. d; Availability is the ratio of actual production time divided by planned production time. The 30-minute lunch shutdown is planned. [VII.B.4]

$$\text{Availability} = \frac{\text{Actual operating time}}{\text{Planned operating time}} = \frac{(8 \times 60) - 30 - 12}{(8 \times 60) - 30} = \frac{438}{450} = 0.9733$$

137. c; If we do not have any information on the percentage defective, we use $p = 0.50$ in the calculation since it results in the most conservative sample size. Here, we use

$$n \geq \frac{z^2 p(1-p)}{E^2} = \frac{1.96^2 \times 0.25}{0.03^2} = 1067.11$$

$$n = 1068.$$

[VI.B.4]

138. a; The evaluation method that uses a three-dimensional miniature version of a solution is known as a physical model. [VII.C]

139. b; The coefficient in the coded regression model for a full factorial experiment will be one-half the factor effect. Here, 5.72 ÷ 2 = 2.86. [VII.A.6]

140. b; In a balanced scorecard, the learning and growth perspective includes the capabilities and skills of an organization. [II.C.1]

141. d; If an experiment must be conducted over two days, the analyst should use blocking to account for any additional variability introduced by different conditions day to day. [VII.A.2]

142. a; Two-factor authentication is an example of mistake-proofing. [VII.B.1]

143. b; A typical error that a leader can make when managing change is to assign the change-management function of people without the resources to execute it. [I.B.2]

144. d; For a minimum tolerance (e.g., diameter > 2.1 cm), the larger-the-better loss function would be used. [IX.C]

145. d; A kaizen blitz is best used when improvement opportunities are readily identifiable, when the risk of failure is small, and when projects are time-bound and clearly defined. [VII.B.3]

146. a; Guidelines for Design for Manufacturability include designing in self-diagnostics. [IX.B]

147. a; The steps in the theory of constraints are, in order, identify, exploit, subordinate, elevate, and repeat. [VII.B.4]

148. c; The upper limit of a p-chart changes based on the value of the sample size, n. Larger samples result in tighter limits. [VIII.A.4]

149. b; When analyzing survey data for a net promoter score, customers responding to the question with a score of 7 or 8 are satisfied but unenthusiastic about the product. [II.C.1]

150. d; The fourth stage in a Six Sigma maturity model is institutionalization. The five stages include (1) launch, (2) early success, (3) scale and replication, (4) institutionalization, (5) culture transformation. [I.A.1]

ASQ Six Sigma Black Belt Certification Body of Knowledge (2022)

The topics in this Body of Knowledge include additional detail in the form of subtext explanations and the cognitive level at which test questions will be written. This information will provide guidance for the candidate preparing to take the exam. The subtext is not intended to limit the subject matter or be all-inclusive of what might be covered in an exam. It is meant to clarify the type of content to be included in the exam. The descriptor in parentheses at the end of each entry refers to the maximum cognitive level at which the topic will be tested. A complete description of cognitive levels is provided at the end of this document.

I. Organization-Wide Planning and Deployment (12 Questions)

A. *Organization-wide considerations*

1. *Fundamentals of six sigma and lean methodologies.* Define and describe the value and goals of these approaches and describe the integration and complementary relationship between them. Identify and understand an organization's lean six sigma maturity model. (Understand)

2. *Six sigma, lean, and continuous improvement methodologies.* Describe when to use six sigma instead of other problem-solving approaches and describe the importance of aligning six sigma objectives with organizational goals. Describe screening criteria and how such criteria can be used for the selection of six sigma projects, lean initiatives, and other continuous improvement methods. (Apply)

3. *Relationships among business systems and processes.* Describe the interactive relationships among business systems, processes, and internal and external stakeholders, and the impact those relationships have on business systems. (Understand)

4. *Strategic planning and deployment for initiatives.* Define the importance of identification and strategic planning of six sigma projects and lean initiatives as part of the business planning process taking into consideration outcomes such as return on investment and measured intangibles. Use strengths, weaknesses, opportunities, and threats analysis

(SWOT), contingency planning, and business continuity planning to enhance strategic planning and deployment. (Apply)

B. ***Leadership***

1. *Roles and responsibilities.* Describe the roles and responsibilities of executive leadership, champions, sponsors, process owners, master black belts, black belts, and green belts in driving six sigma and lean initiatives. Understand the importance of coaching multiple levels of leadership and belts on appropriate six sigma tools and techniques. Understand the importance of finance and its role supporting a project and confirming its outcome. Describe how each group influences project deployment in terms of providing or managing resources, enabling changes in organizational structure, and supporting communications about the purpose and deployment of the initiatives. (Understand)

2. *Organizational barriers.* Describe how an organization's structure and culture can impact six sigma projects. Identify common causes of six sigma failures, including lack of management support and lack of resources. (Apply)

3. *Change management.* Apply and facilitate change management techniques, including stakeholder analysis, readiness assessments, proactive change management, and communication plans to overcome barriers and drive organization-wide change. (Apply)

II. Organizational Process Management and Measures (12 Questions)

A. ***Impact on stakeholders.*** Describe the impact six sigma projects can have on customers, suppliers, and other stakeholders. (Understand)

B. ***Benchmarking.*** Define and distinguish between various types of benchmarking, e.g., best practices, competitive, collaborative, and breakthrough. Select measures and performance goals for projects resulting from benchmarking activities. (Apply)

C. ***Business measures***

1. *Performance measures.* Define and describe balanced scorecard, key performance indicators (KPIs), key behavior indicators (KBIs), objectives and key results (OKRs), customer loyalty metrics, and leading and lagging indicators. Explain how to create a line of sight from performance measures to organizational strategies. (Analyze)

2. *Financial measures.* Define and use revenue growth, market share, margin, net present value (NPV), return on investment (ROI), and cost–benefit analysis (CBA). Explain the difference between hard cost measures (from

profit and loss statements) and soft cost benefits of cost avoidance and reduction. (Apply)

III. Team Management (15 Questions)

A. *Team formation*

1. *Team types and constraints.* Define and describe various teams, including virtual, cross-functional, and self-directed. Determine what team type will work best for a given set of constraints, e.g., geography, technology availability, staff schedules, time zones. (Apply)

2. *Team roles and responsibilities.* Define and describe various team roles and responsibilities for leader, facilitator, coach, and individual member. (Understand)

3. *Team member selection criteria.* Describe various factors that influence the selection of team members, including the ability to influence, openness to change, required skills sets, subject matter expertise, and availability. (Apply)

4. *Team success factors.* Identify and describe the elements necessary for successful teams, e.g., management support, clear goals, ground rules, and timelines. (Apply)

B. *Team facilitation*

1. *Motivational techniques.* Describe and apply techniques to motivate team members. Identify factors that can demotivate team members and describe techniques to overcome them. (Apply)

2. *Team stages of development.* Identify and describe the classic stages of team development: forming, storming, norming, performing, and adjourning. (Apply)

3. *Team communication.* Describe and explain the elements of an effective communication plan, e.g., audience identification, message type, medium, frequency. (Apply)

4. *Team leadership models.* Describe and select appropriate leadership approaches (e.g., direct, coach, support, and delegate) to ensure team success. (Apply)

C. *Team dynamics*

1. *Group behaviors.* Identify and use various conflict resolution techniques (e.g., coaching, mentoring, and intervention) to overcome negative group dynamics, including dominant and reluctant participants, groupthink, rushing to finish, and digressions. (Evaluate)

2. *Meeting management.* Select and use various meeting management techniques, including using agendas, starting on time, requiring pre-work by attendees, and ensuring that the right people and resources are available. (Apply)

3. *Team decision-making methods.* Define, select, and use various tools (e.g., consensus, nominal group technique, and multi-voting) for decision making. (Apply)

D. *Team training*

1. *Needs assessment.* Identify the steps involved to implement an effective training curriculum: identify skills gaps, develop learning objectives, prepare a training plan, and develop training materials. (Understand)

2. *Delivery.* Describe various techniques used to deliver effective training, including adult learning theory, soft skills, and modes of learning. (Understand)

3. *Evaluation.* Describe various techniques to evaluate training, including evaluation planning, feedback surveys, pre-training and post-training testing. (Understand)

IV. Define (20 Questions)

A. *Voice of the customer*

1. *Customer identification.* Identify and segment customers and show how a project will impact both internal and external customers. (Apply)

2. *Customer data collection.* Identify and select appropriate data collection methods (e.g., surveys, focus groups, interviews, and observations) to gather voice of the customer data. Ensure the data collection methods used are reviewed for validity and reliability. (Analyze)

3. *Customer requirements.* Define, select, and apply appropriate tools to determine customer needs and requirements, including critical-to-X (CTX when "X" can be quality, cost, safety, etc.), CTQ tree, quality function deployment (QFD), supplier, input, process, output, customer (SIPOC), and Kano model. (Analyze)

B. *Business case and project charter*

1. *Business case.* Describe business case justification used to support projects. (Understand)

2. *Problem statement.* Develop a project problem statement and evaluate it in relation to baseline performance and improvement goals. (Evaluate)

3. *Project scope.* Develop and review project boundaries to ensure that the project has value to the customer. (Analyze)

4. *Goals and objectives.* Identify SMART (specific, measurable, actionable, relevant, and time-bound) goals and objectives on the basis of the project's problem statement and scope. (Analyze)

5. *Project performance measurements.* Identify and evaluate performance measurements (e.g., cost, revenue, delivery, schedule, and customer satisfaction) that connect critical elements of the process to key outputs. (Analyze)

6. *Project charter review.* Explain the importance of having periodic project charter reviews with stakeholders. (Understand)

C. ***Project management (PM) tools.*** Identify and use the following PM tools to track projects and document their progress. (Evaluate)

1. Gantt charts
2. Tollgate reviews
3. Work breakdown structure (WBS)
4. RACI Model (responsible, accountable, consulted, and informed)

D. ***Analytical tools.*** Identify and use the following analytical tools throughout the DMAIC cycle. (Apply)

1. Affinity diagrams
2. Tree diagrams
3. Matrix diagrams
4. Prioritization matrices
5. Activity network diagrams
6. Process decision program chart (PDPC)
7. Interrelationship digraph (ID)

V. Measure (25 Questions)

A. ***Process characteristics***

1. *Process flow metrics.* Identify and use process flow metrics (e.g., work in progress (WIP), work in queue (WIQ), touch time, takt time, cycle time, and throughput) to determine constraints. Describe the impact that "hidden factories" can have on process flow metrics. (Analyze)

2. *Process analysis tools*. Select, use, and evaluate various tools, e.g., value stream maps, process maps, work instructions, flowcharts, spaghetti diagrams, and gemba walk. (Evaluate)

B. *Data collection*

1. *Types of data*. Define, classify, and distinguish between qualitative and quantitative data, and continuous and discrete data. (Evaluate)

2. *Measurement scales*. Define and use nominal, ordinal, interval, and ratio measurement scales. (Apply)

3. *Sampling*. Define and describe sampling concepts, including representative selection, homogeneity, bias, accuracy, and precision. Determine the appropriate sampling method (e.g., random, stratified, systematic, subgroup, and block) to obtain valid representation in various situations. (Evaluate)

4. *Data collection plans and methods*. Develop and implement data collection plans that include data integrity, accuracy, and processing tools, e.g., check sheets and data normalization. Avoid data collection pitfalls by defining the metrics to be used or collected, ensuring that collectors are trained in the tools and understand how the data will be used, and checking for seasonality effects. (Analyze)

C. *Measurement systems*

1. *Measurement system analysis (MSA)*. Use gauge repeatability and reproducibility (R&R) studies and other MSA tools (e.g., bias, correlation, linearity, precision to tolerance, and percent agreement) to analyze variable measurement system capability. Use audit MSA for attribute measurement systems. (Evaluate)

2. *Measurement systems across the organization*. Identify how measurement systems can be applied to all functional areas of the organization (e.g., marketing, sales, engineering, research and development (R&D), supply chain management, operations and customer experience). (Understand)

3. *Metrology*. Define and describe elements of metrology, including calibration systems, traceability to reference standards, and the control and integrity of measurement devices and standards. (Understand)

D. *Basic statistics*

1. *Basic statistical terms*. Define and distinguish between population parameters and sample statistics, e.g., proportion, mean, and standard deviation. (Apply)

2. *Central limit theorem.* Explain the central limit theorem and its significance in the application of inferential statistics for confidence intervals, hypothesis tests, and control charts. (Understand)
3. *Descriptive statistics.* Calculate and interpret measures of dispersion and central tendency. (Evaluate)
4. *Graphical methods.* Construct and interpret diagrams and charts, e.g., box-and-whisker plots, scatter diagrams, histograms, normal probability plots, frequency distributions, and cumulative frequency distributions. (Evaluate)
5. *Valid statistical conclusions.* Distinguish between descriptive and inferential statistical studies. Evaluate how the results of statistical studies are used to draw valid conclusions. (Evaluate)

E. *Probability*

1. *Basic concepts.* Describe and apply probability concepts, e.g., independence, mutually exclusive events, addition and multiplication rules, conditional probability, complementary probability, and joint occurrence of events. (Understand)
2. *Common distributions.* Describe, interpret, and use normal, Poisson, binomial, chi-square, Student's t, and F distributions. (Evaluate)
3. *Additional distributions.* Identify hypergeometric, bivariate, exponential, lognormal, and Weibull distributions. (Understand)

F. *Process capability*

1. *Process capability indices.* Define, select, and calculate C_p and C_{pk}. (Evaluate)
2. *Process performance indices.* Define, select, and calculate P_p, P_{pk}, C_{pm}, and process sigma. (Evaluate)
3. *General process capability studies.* Describe and apply elements of designing and conducting process capability studies relative to characteristics, specifications, sampling plans, stability, and normality. (Evaluate)
4. *Process capability for attributes data.* Calculate the process capability and process sigma level for attributes data. (Apply)
5. *Process capability for non-normal data.* Identify non-normal data and determine when it is appropriate to use Box-Cox or other transformation techniques. (Apply)
6. *Process performance versus specification.* Distinguish between natural process limits and specification limits. Calculate process performance metrics,

e.g., percent defective, parts per million (PPM), defects per million opportunities (DPMO), defects per unit (DPU), first pass yield, and rolled throughput yield (RTY). (Evaluate)

7. *Short-term and long-term capability*. Describe and use appropriate assumptions and conventions when only short-term data or only long-term data are available. Interpret the relationship between short-term and long-term capability. (Evaluate)

VI. Analyze (22 Questions)

A. *Measuring and modeling relationships and variables*

1. *Correlation coefficient*. Calculate and interpret the correlation coefficient and its confidence interval and describe the difference between correlation and causation. (Evaluate)

2. *Linear regression*. Calculate and interpret regression analysis and apply and interpret hypothesis tests for regression parameters. Use the regression model for estimation and prediction, analyze the uncertainty in the estimate, and perform a residual analysis to validate the model. (Evaluate)

3. *Multivariate tools*. Understand sources of variation through multivariate (e.g., factor analysis, discriminant analysis, multiple analysis of variance (MANOVA)). (Understand)

B. *Hypothesis testing*

1. *Terminology*. Define and interpret the significance level, power, type I, and type II errors of statistical tests. (Evaluate)

2. *Statistical versus practical significance*. Define, compare, and interpret statistical and practical significance. (Evaluate)

3. *Sample size*. Calculate sample size for common hypothesis tests: equality of means and equality of proportions. (Apply)

4. *Point and interval estimates*. Define and distinguish between confidence and prediction intervals. Define and interpret the efficiency and bias of estimators. Calculate tolerance and confidence intervals. (Evaluate)

5. *Tests for means, variances, and proportions*. Use and interpret the results of hypothesis tests for means, variances, and proportions. (Evaluate)

6. *Analysis of variance (ANOVA)*. Select, calculate, and interpret the results of ANOVAs. (Evaluate)

7. *Goodness-of-fit (chi-square) tests*. Define, select, and interpret the results of these tests. (Evaluate)

C. *Risk analysis and management*

1. *Types of risk.* Identify, assess, and prioritize various types of risk such as enterprise, operational, supplier, security, product, and cybersecurity. (Analyze)

2. *Failure mode and effects analysis (FMEA).* Describe the purpose and elements of FMEA, including risk priority number (RPN), and evaluate FMEA results for processes, products, and services. Distinguish between design FMEA (DFMEA) and process FMEA (PFMEA) and interpret their results. (Evaluate)

D. *Additional analysis methods*

1. *Gap analysis.* Analyze scenarios to identify performance gaps, and compare current and future states using predefined metrics. (Analyze)

2. *Root cause analysis.* Define and describe the purpose of root cause analysis, recognize the issues involved in identifying a root cause, and use various tools (e.g., 5 whys, Pareto charts, fault tree analysis, cause-and-effect diagrams, and A3) to resolve chronic problems. (Analyze)

3. *Waste analysis.* Identify and interpret the seven classic wastes (overproduction, inventory, defects, over-processing, waiting, motion, and transportation) and resource under-utilization. (Analyze)

VII. Improve (21 Questions)

A. *Design of experiments (DOE)*

1. *Terminology.* Define basic DOE terms, e.g., independent and dependent variables, factors and levels, response, treatment, error, and nested. (Understand)

2. *Design principles.* Define and apply DOE principles, e.g., power, sample size, balance, repetition, replication, order, efficiency, randomization, blocking, interaction, confounding, and resolution. (Apply)

3. *Planning experiments.* Plan and evaluate DOEs by determining the objective, selecting appropriate factors, responses, and measurement methods, and choosing the appropriate design. (Evaluate)

4. *One-factor experiments.* Understand when to use completely randomized, randomized block, and Latin square designs. (Understand)

5. *Two-level fractional factorial experiments.* Understand these types of experiments and describe how confounding can affect their use. (Understand)

6. *Full-factorial experiments*. Understand these types of experiments. (Understand)

B. ***Lean methods***

1. *Waste elimination*. Select and apply tools and techniques for eliminating or preventing waste, e.g., pull systems, kanban, 5S, standard work, and poka-yoke. (Analyze)

2. *Cycle-time reduction*. Use various tools and techniques for reducing cycle time, e.g., continuous flow, single-minute exchange of die (SMED), and heijunka (production leveling). (Analyze)

3. *Kaizen*. Define and distinguish between kaizen and kaizen blitz and describe when to use each method. (Apply)

4. *Other improvement tools and techniques*. Identify and describe how other process improvement methodologies are used, e.g., theory of constraints (TOC), and overall equipment effectiveness (OEE). (Understand)

C. ***Implementation.*** Develop plans for implementing proposed improvements, including conducting pilot tests or simulations, and evaluate results to select the optimum solution. (Evaluate)

VIII. Control (17 Questions)

A. ***Statistical process control (SPC)***

1. *Objectives*. Explain the objectives of SPC, including monitoring and controlling process performance, tracking trends, runs, and reducing variation within a process. (Understand)

2. *Selection of variables*. Identify and select critical process characteristics for control chart monitoring. (Apply)

3. *Rational subgrouping*. Define and apply the principle of rational subgrouping. (Apply)

4. *Control chart selection*. Select and use control charts in various situations: $\bar{X}-R$, $\bar{X}-s$, individual and moving range (ImR), p, np, c, u, short-run SPC, and moving average. (Apply)

5. *Control chart analysis*. Interpret control charts and distinguish between common and special causes using rules for determining statistical control. (Analyze)

B. *Other controls*

1. *Total productive maintenance (TPM).* Define the elements of TPM and describe how it can be used to consistently control the improved process. (Understand)

2. *Visual controls.* Define the elements of visual controls (e.g., pictures of correct procedures, color-coded components, and indicator lights), and describe how they can help control the improved process. (Understand)

C. *Maintain controls*

1. *Measurement system reanalysis.* Review and evaluate measurement system capability as process capability improves, and ensure that measurement capability is sufficient for its intended use. (Evaluate)

2. *Control plan.* Develop a control plan to maintain the improved process performance, enable continuous improvement, and transfer responsibility from the project team to the process owner. (Apply)

D. *Sustain improvements*

1. *Lessons learned.* Document the lessons learned and benefits realized from all phases of a project and identify strategies for reinforcing and replicating improvements. (Apply)

2. *Documentation.* Develop or modify documents including standard operating procedures (SOPs), work instructions, and control plans to ensure that the improvements are sustained over time. (Apply)

3. *Training for process owners and staff.* Develop and implement training plans that are handed off to process owners to ensure consistent execution of revised process methods, KPIs to confirm sustained benefits, and standards to maintain process requirements. (Apply)

4. *Ongoing evaluation.* Identify and apply tools (e.g., control charts and control plans) for ongoing evaluation of the improved process, including monitoring leading indicators, lagging indicators, and additional opportunities for improvement. (Apply)

Design for Six Sigma (DFSS) Framework and Methodologies (6 Questions)

A. *Common DFSS methodologies.* Identify and describe DMADV (define, measure, analyze, design, and validate) and DMADOV (define, measure, analyze, design, optimize, and validate). (Understand)

B. *Design for X (DFX).* Describe design constraints, including Design for cost, Design for manufacturability (producibility), Design for test, and Design for maintainability. (Understand)

C. ***Robust designs.*** Describe the elements of robust product design, tolerance design, and statistical tolerancing. (Understand)

LEVELS OF COGNITION—BASED ON BLOOM'S TAXONOMY—REVISED (2001)

In addition to content specifics, the subtext for each topic in this BoK also indicates the intended complexity level of the test questions for that topic. These levels are based on "Levels of Cognition" (from Bloom's Taxonomy—Revised, 2001) and are presented below in rank order, from least complex to most complex.

Remember

Recall or recognize terms, definitions, facts, ideas, materials, patterns, sequences, methods, principles, etc.

Understand

Read and understand descriptions, communications, reports, tables, diagrams, directions, regulations, etc.

Apply

Know when and how to use ideas, procedures, methods, formulas, principles, theories, etc.

Analyze

Break down information into its constituent parts and recognize their relationship to one another and how they are organized; identify sublevel factors or salient data from a complex scenario.

Evaluate

Make judgments about the value of proposed ideas, solutions, etc., by comparing the proposal to specific criteria or standards.

Create

Put parts or elements together in such a way as to reveal a pattern or structure not clearly there before; identify which data or information from a complex set is appropriate to examine further or from which supported conclusions can be drawn.

References

Borror, Connie M., and Sarah E. Burke, 2019. *The ASQ CQE Study Guide.* Milwaukee: ASQ Quality Press.

Durivage, Mark Allen, and Shawn Findlater, 2018. *The ASQ CSSBB Study Guide.* Milwaukee: ASQ Quality Press.

McShane-Vaughn, Mary. 2016. *The Probability Handbook.* Milwaukee: ASQ Quality Press.

———. 2021. *Lean Six Sigma Leadership Tools for Black Belts.* Folsom, CA: University Training Partners Publishing.

———. 2021. *Statistics for Black Belts Slide Book.* Folsom, CA: University Training Partners Publishing.

———. ed. 2022. *The ASQ Certified Six Sigma Black Belt Handbook.* Milwaukee: ASQ Quality Press.

McShane-Vaughn, Mary, and C. Grant Short. 2023. *The Handbook of Quality Equations.* Folsom, CA: University Training Partners Publishing.

Montgomery, Douglas C. 2019. *Design and Analysis of Experiments,* 10th ed. Hoboken, NJ: John Wiley & Sons.

———. 2005. *Introduction to Statistical Quality Control,* 5th ed. Hoboken, NJ: John Wiley.

Montgomery, Douglas C., and George C. Runger. 2010. *Applied Statistics and Probability for Engineers,* 5th ed. Hoboken, NJ: John Wiley.

Tague, Nancy R. 2005. *The Quality Toolbox,* 2nd ed. Milwaukee: ASQ Quality Press.

Triola, Mario F. 2006. *Elementary Statistics,* 10th ed. Boston: Pearson Education.

www.ingramcontent.com/pod-product-compliance
Lightning Source LLC
LaVergne TN
LVHW060029210926
848481LV00013B/3820
* 9 7 8 1 6 3 6 9 4 1 1 6 5 *